COWPEA IN INDIA

OTHER BOOKS BY SAME EDITORS

GUAR IN INDIA

Guar in recent years has achieved status of an industrial crop due to water soluble natural polymer galactomannan gum, used for versatile industrial purposes. The crop has great scope as foreign exchange earner and guar gum worth Rs. 1000/- crores may be annually exported. It is, therefore, in national interest that production of guar and it's gum are sustained under the prevailing rainfed situations.

Technically the production of guar would be sustained if latest available know-how is compiled and made available to the intended consumers, growers and exporters. It would expose bottlenecks in the technological advancements on their implementation on the one, and their effective utilization on the other. A compilation on the latest research and developmental aspects aiming at boosting grain production, export of gum and seed meal, is therefore, judiciously required.

The book chapters have been devoted to genetic improvement, agronomy, genetic resources, fodder aspects, microbiology, physiology, pathology, entomology, gum quality, industrial aspects and seed production of guar. The contents of the book are comprehensive and convenient which may prove useful in academic and applied context.

ISBN: 81-7233-310-2

MOTH BEAN IN INDIA

Moth bean is an important crop for arid areas. It has multi-uses and adapted to extremes of uncongenial ecological niches particularly, in areas receiving less rains with erratic distribution. The crop is credited with a number of adaptive morpho-physiological features. It however, requires genetic improvement in respect of efficient plant types, agronomic improvement, plant protection cares. There is also need on assessment of genetic resources available, quality considerations. Information on these aspects is scarce and unsystematically arranged. Efforts on compilation of available research results would pave a way for furthering moth bean in required aspects and fields. Efforts have, therefore, been made to collect as far as possible vast information covering past 50 years on moth bean related to different aspects of adaptation, genetic resources, genetic and agronomic improvements, plant protection strategies, biotechnological possibilities and quality considerations.

The book, it is hoped, would prove useful and informative to those concerned with moth bean in one respect or other.

ISBN: 81-7233-311-0

COWPEA IN INDIA

D. KUMAR
N.B. SINGH

SCIENTIFIC PUBLISHERS (INDIA)
P.O. BOX 91 JODHPUR

Published on behalf of
INDIAN ARID LEGUMES SOCIETY

Published by:
PAWAN KUMAR
Scientific Publishers (India)
5-A, New Pali Road, P.O. Box 91
JODHPUR – 342 001
Tel.: +91-291-2433323
Fax.: +91-291-2512580
E-mail: info@scientificpub.com
www.scientificpub.com

Print : 2021

ISBN: 81-7233-367-6

Lasertype Set: Rajesh Ojha
Printed in India

FOREWORD

Cowpea, an ancient pulse, is characteristically adapted to dry habitats and fragile agro-eco systems of tropical and sub-tropical regions in Asia, Africa, Central-Southern America and parts of southern USA and Europe. It is considered as an important food legume for livelihood of millions of rural masses in the underdeveloped third world. This legume has great potential in India for successful cultivation in kharif and summer in northern India and throughout the year in peninsular regions. An important component of farming systems in resource constraint agriculture, it is vitally used for food, feed, forage, vegetable, green manuring purposes and grains are used for several snacks. Besides, providing substantial vegetable protein, cowpea is also a good source of essential amino acids viz., lysine, leucine and phenylalanine and minerals like calcium and iron. Initial fast growth of cowpea with broad and dropping leaves provides a rich soil coverage therefore, reducing soil temperature and conserving the soil during warm months.

Cowpea known for several adapted merits and common men's uses, suffers from inherited defects and demerits of loose, tall, trailing-viny, indeterminate growth habits and late partitioning with poor conversion of large dry matter towards economic parts. Cowpea therefore, represents long repeated natural selections, resulting in poor productivity. Concerted and renewed efforts are, therefore, needed to alter its plant type which is more productive and responsive to agronomic inputs imparted with inbuilt tolerance to specific biotic and abiotic constraints.

There is, therefore, need to collect and compile research results related to various production components, being generated through individuals, institutions and coordinated project and publish in condensed form so that relevant research results are used by the intented scientists, students and field workers. I applaud sincere efforts undertaken by the authors and congratulate them for bringing out present compilation **"Cowpea in India"** at the needed hour.

It is believed that present compilation would be of great use and help to all those engaged and interested in integrated development and advancement of cowpea in India.

05-12-03

Dt. 05.12.2003

(A.S. Faroda)
Chairman,
Agricultural Scientists' Recruitment Board, ICAR
New Delhi

PREFACE

The national agricultural research has made significant achievements and strides in almost all the major food sectors. For instance, food grain productivity just stagnating to 522 kg ha-1 during 1950 appreciably surged to 1723 kg ha^{-1} during 2001, registering a remarkable gain of almost 3.3 folds, similarly productivity of other food commodities viz., fruits, milk, vegetables, fishries and eggs have spurted to the tune of almost 1.6, 1.8, 2.1, 5.6 and 6.4 folds during this period. These achievements have nailed the continued success of Indian agricultural research and development efforts at the national and international levels. However, the cause of great concern and worry to us all is the sluggish and hovering contribution of pulses to the food grain basket. For instance, the proportional gap between cereals and pulses that was 5:1 during 1950 considerably widened to 11.5:1 during late nineties. Priorities of Indian agriculture R&D are therefore, concentrating at pulse production so as to provide nutritious balanced diets to majority of the inhabitants relishing on vegetarian food habits.

There can be several reasons for poor productivity of pulses including technological, social, financial, geographical, climatological etc., but important is delayed and slow technological breakthrough particularly, in seed based improvement that could have far reaching acceptance and adoption across the climatic zones and the farming communities.

Of more than two and half a dozen pulses, being grown in India, cowpea commonly known as *lobia* is one of principal ancient pulses. It is characteristically known for wide geographic distribution across the soil, soil inputs and climatological zones. Hence, it is widely grown in tropics and sub-tropics of Asia, Africa, Central and Southern America and parts of Southern Europe and U.S.A. Central and Western Africa only account for more than 60% of cowpea area in the world. However, substantial cowpea production is realized from to drier regions of Nigeria. Cowpea growing countries in Asia are : India, Sri Lanka, Bangladesh, Myanmar, China, Korea, Thailand, Indonesia, Malaysia, Pakistan and Nepal. It is estimated to be cultivated in almost 12.5 million hectares of lands with annual production of almost 3.0 million tons worldwide.

In India, cowpea is grown as sole, inter-crop, mix-crop and in agro-forestry combinations, hence, exact statistics on its area are not available but is estimated to be cultivated in almost half of 1.3 million hectares of area occupied by Asian region. Cowpea considred as a minor pulse in India context is cultivated in aird and semi-arid tracts of Rajasthan, Gujarat, Karnataka, Tamilnadu, Maharashtra and in traces in majority of other states.

Cowpea is known for drought hardy nature, its wide and droppy leaves keep soil and soil moisture conserved due to shading effects. Initial fast growth with fast penetrating root system and strong stomatal sensitivity justify its initial establishment in soil moisture deficit situations. Cowpea has multiple uses as food, feed, forage, fodder, vegetable and is a good source of protein, calcium and iron.

In spite of a series of adaptive and livelihood uses, cowpea is known for poor partitioning of large dry matter production towards pods formation, precisely it is characterized with poor source-sink relationship. It suffers from loose, spreading, viny-trailing plant type with delayed partitioning and maturity. The crop therefore, appears to have been evolved for survival importance and little for productive value. Besides, poor and less productive plant type, crop also suffers from serious fungal and bacterial diseases like, Anthracnose, Charcoal rot, Cercospora leaf spot, Bacterial blight and certain nematodes and viral diseases like, Root knot, CYMV etc, and insect-pests like, pod borer and bruchid species. Crop also suffers from poor agronomic response and susceptibility to shattering and salt adversity.

In view to imporve cowpea for certain defects and bottlenecks, serious efforts have been undergoing at various institutions in individual capacity and coordinated manner generating vital technological information. The research results so generated during, past several years are still at large of technocrats, planners and those engaged in extension of new technological advancements to the grass root users.

In view of above, it would be quite appropriate to collect, compile and published the condensed research results of scholary and practical relevance concerning cowpea pertaining to last five decades on plant improvement, genetics, genetic resources, its husbandry, biotic and abiotic aspects under one umbrella. The present compilation **"Cowpea in India"** represents first ever attempt in India as far cowpea research on aforesaid aspects are concerned.

It is believed that present compilation would be of great help and use to those engaged in all round cowpea improvement and development in India.

D. Kumar **N.B. Singh**

CONTRIBUTORS

1. D. Kumar,
Project Coordinator – Arid Legumes,
Central Arid Zone Research Institute,
Jodhpur – 342 003, Rajasthan

2. H.R. Mahla,
Research Fellow,
Central Arid Zone Research Institute,
Jodhpur – 342 003, Rajasthan

3. R.S. Sangwan,
Assoc. Professor (Pl. Breeding)
Department of Plant Breeding,
Choudhary Charan Singh Haryana Agricultural University,
Hisar – 125 004, Haryana

4. S.K. Mishra,
Senior Scientist (Plant Breeding),
Division of Genetics,
Indian Agricultural Research Institute, **New Delhi** – 110 012

5. D.D. Saini,
Associate Professor (Plant Breeding),
Rajasthan Agricultural University, Regional Research Station,
Durgapura, **Jaipur**-302 018, Rajasthan

6. S.P.S. Chaudhary,
Associate Professor (Plant Breeding),
Rajasthan Agricultural University, Regional Research Station,
Durgapura, **Jaipur** -302 018, Rajasthan

7. R.N. Pandey,
Scientific Officer (E),
Nuclear Agriculture and Biotechnology Division,
Bhaba Atomic Research Centre, Trombay,
Mumbai – 400 085, Maharashtra

8. S.E. Pawar,
Scientific Officer E (Retd.),
Nuclear Agriculture and Biotechnology Division,
Bhaba Atomic Research Centre, Trombay,
Mumbai – 400 085, Maharashtra

9. S. Anitha,
Assistant Agronomist,
Kerala Agricultural University, Regional Research Station,
Pattambi-679 306, Kerala

10. E. Sreenivasan,
Professor (Botany),
Kerala Agricultural University Regional Research Station,
Pattambi-679 306, Kerala

12. P.P. Gupta,
Professor (Plant Pathology),Forage Section,
Chaudhary Charan Singh Haryana Agricultural University,
Hisar – 125 004, Haryana

13. Vivek Gupta,
Research Fellow, Forage Section,
Chaudhary Charan Singh Haryana Agricultural University,
Hisar – 125 004, Haryana

14. Satya Vir,
Principal Scientist (Entomology),
Central Arid Zone Research Institute,
Jodhpur – 342 003, Rajasthan

15. R.K. Kaul,
Senior Scientist (Nematology)
Central Arid Zone Research Institute,
Jodhpur – 342 003, Rajasthan

16. C.R. Ramesh,
Principal Scientist and Head, IGFRI,
Regional Research Station,
U.A.S. Campus, **Dharwad** – 580 005, Karnatka

17. P.S. Pathak,
Director,
Indian Grassland and Fodder Research Institute,
Jhansi – 284 003, U.P.

18. B. Gangaiah,
Sr. Scientist (Agronomy),
Indian Agricultural Research Institute,
New Delhi – 110 012

19. Pratibha Tewari,
Sr. Scientist (Home Science),
Central Arid Zone Research Institute,
Jodhpur – 342 003, Rajasthan

20. Sanju Choudhary,
Senior Research Fellow,
Central Arid Zone Research Institute,
Jodhpur – 342 003, Rajasthan

21. Nidhi Shekhawat,
Senior Research Fellow,
Central Arid Zone Research Institute,
Jodhpur – 342 003, Rajasthan

22. N.B.Gaddagimath,
Director, R&D,
Sarpan Agri Horticulture Research Centre,
Dharwad-5, Karnataka

CONTENTS

CHAPTER – 1

COWPEA - AN INTRODUCTION

D. Kumar and H.R. Mahla

Cowpea [*Vigna unguiculata* (L.) Walp] one of the oldest source of human food, has most likely been used as a crop since, Neolithic times. Cowpea is known by a number of common names, viz., *crowdel pea, black-eyed pea, southern pea* and internationally as *lobia, niebe, coupe* or *frijole*. An annual legume of ancient origin, it is cultivated world over under varied soil and rainfall situations in tropics and subtropic regions of Asia, Africa, Central and Southern America, parts of Southern Europe and USA (Mishra *et al.*, 1985; Singh and Natore, 1985; Fery, 1990). Cowpea growing countries in the Asian region are India, Sri Lanka, Bangladesh, Myanmar, China, Korea, Thiland, Indonesia, Nepal, Pakistan, Philippines and Malaysia. India alone accounts almost half of the total Asian acreage. It is difficult to estimate the precise statistics on area and production of cowpea because of the fact, same is grown as mixed and intercrop. In India, it is mainly grown in semi-arid regions, however, Western, Central and peninsular regions are the major contributors.

The views on origin of cowpea are contradictory, revealing Africa, Asia and South America among the probable regions of its origin. The Oldest view is that cowpea was probably introduced from Africa to the Indian sub-continent approximately 2000 to 3500 years ago. Before 300 BC, cowpea had reached Europe and most likely to North Africa from Asian regions. During 17th centaury AD, the Spanish collected the crop to the West Indies. Padulosi and Ng (1997) speculated that the Transvaal region of the Republic of South Africa could be the centre of speciation of *V. unguiculata*, due to prevalence of most primitive wild tribes. It was further hypothesized that this species moved towards northern regions of Transvaal to Mozambique and Tanzania, where the sub-species pubescens was evolved.

Marginal and sub marginal farmers in the semi-arid and sub humid regions of Africa have been major producers and consumers of cowpeas. Steele *et al.*, (1985) noted that the protein content of the leafy cowpea parts consumed annually in Africa and Asia is equivalent of 5 million tonnes of dry cowpea seeds and that this represents as much as 30% of the total food legumes production in the lowland tropics. According to FAO estimate, 3.3 million tones of dry grains of cowpea were produced worldover in 2000, of the same Nigeria alone produced 2.1 million tones making it, therefore, world's largest producer, followed by Niger (0.65 mt) and Mali (0.11 mt). Total global area grown to cowpea was estimated to 9.8 million hectares, of the same about 9.3 million hectares was occupied by West Africa. World cowpea average yield was 337 kg per hectare, whereas productivity of Nigeria was higher to the tune of 417 kg per hectare.

Botany

Cowpea plants are often categorized having erect, semi-erect, prostrate (trailing), or climbing types with considerable variability within species. Growth habit may be represented from indeterminate to fairly determinate with non-viny types tending to be more determinate. Cowpea is strongly a tap rooted annual legume. Root depth may be measured to as 95 inches almost 8 weeks of seeding. Emergence is epigeal (similar to common bean, and lupin) where the cotyledons emerge out from the ground during germination. The trifoliate leaves develop alternately are smooth, dull to shiny, and rarely pubescent. Commonly, the terminal leaflet is longer and larger than the lateral ones. There is a wide range in leaf size and its shape. Flowers are borne in multiple racemes on 8 to 20 flower stalks (peduncles), arising from the leaf axil. Flowers open early in the morning, close by noon and fall the same day. The open display of flowers above the foliage and the presence of floral nectaries contribute to the attraction of insects. Two or three pods are produced per peduncle, of these long peduncle is a distinguishing feature of cowpea and this characteristic also facilitates the harvest. Cowpea pods are smooth, 6 to 10 inches long, cylindrical and generally somewhat curved. Seed shape is a major characteristic correlated with its development in the pod. Seeds develop in a kidney shape if unrestricted in the pod. The seed coat may be smooth or wrinkled with variety of colors including white, cream, green, buff red, brown, and black. Seeds may also be speckled, mottled, or blotchy. Seeds may also be referred to as "eyed" (blackeye, pinkeye purple hull, etc.), where the white colored hilum is surrounded by another color. Cowpea primarily is self pollinating type but somewhat out crossing (10-15%) also occurs probably with ants, flies and bees as main vectors.

Species

Cowpea [*Vigna unguiculata* (L.) Walp] belongs to the family leguminoceae, sub family papilionaceae, having chromosome number, 2n = 22. Most taxonomists agree that cowpeas belong to the botanical species *Vigna unguiculata* (L.) Walp. However, classification and nomenclature of taxa at the intra-specific level are still debated. Verdcourt (1970) subdivided the species into 3 subspecies, i.e., cowpea (subspecies *ungui-culata*), catjang (sub species *catjang*), and yardlong bean (subspecies *sesquipedalis*). However, Marechal *et al.*, (1978) reclassified the subspecies *unguiculata*, *catjang*, and *sesquipe-dalis* as cultigroups *unguiculata*, *biflora* and *sesquipedalis*, respectively, and clumped these cultigroups under *V. unguicu-lata* subspecies *unguiculata*. Most cowpea breeders seem to have adopted Marechal *et al.*, (1978) cultigroup scheme for classification of cultivated *V. unguiculata* taxa. Westphal (1974) however, classified it in four distinct cultivar groups viz., *unguiculata*, *biflora* (cylindrical), *sesquipedalis* and *textilis*.

Cowpea contains three cultivated (*unguiculata*, *cylindria* (biflora) and *sesquipedalis*) and two wild species (*dekindtiana* and *mensis*). The sub species and varieties of *unguiculata* are outlined below:

V. unguiculata (L.) Walpers.

ssp. *Biflora*, *sesquipedalis*, textiles

ssp. *tentuis*

ssp. *stenophylla*
ssp. *dekidtiana*
var. *mensis*
var. *pubescens*
var. *protracta*

Uses

Cowpea seed is a nutritious component in the human diet, and cheap livestock feed, as well. Cowpea grown to maturity can be used as a feed, fodder or its pods can be harvested and eaten as a vegetable. The tender green leaves are an important food source. These are mainly used as vegetable, especially by small scale farmers in rural areas. Choice of cowpea as vegetable is due to being palatable, highly nutritious and relatively free of metabolites or other toxins.

The chemical composition of cowpea seeds (Table 1) corresponds to that of most common edible legumes grown in arid and semi-arid regions. The seeds also contain small proportions of β-carotene, thiamin, riboflavin, vitamin A, niacin, folic acid and ascorbic acid. The use of cowpea seeds as a vegetable provides cheap source of protein in vegetarian dominated diets of undeveloped nations. Fresh leaves and fast growing twigs of cowpea are often picked up and eaten like spinach.

Table 1. Chemical composition (%) of different parts of cowpea

S. No.	Constituents	Seed	Hay	Leaves
1.	Carbohydrate	56-66	—	8.0
2.	Protein	22-24	18.0	4.7
3.	Water	11.0	9.6	85.0
4.	Crude fibre	5.9-7.3	23.3	2.0
5.	Ash	3.4-3.9	11.3	—
6.	Fat	1.3-1.5	2.6	0.3
7.	Phosphorous	0.146	—	0.063
8.	Calcium	0.104-0.076	—	0.256
9.	Iron	0.005	—	0.005

Source: Ricardo, 1985

Immature snapped pods are used in the same way as snap beans, often being mixed with other foods. Green cowpea seeds are boiled as a fresh vegetable or may be canned or frozen. Dry mature seeds are also suitable for boiling and canning (Davis *et al.*, 1991).

The beans are nutritious and provide complementary proteins to cereals diets. Some people eat both the fresh pods and leaves, and the dried seeds are popular ingredients in a variety of dishes. It has been grouped into the following market classes based on seed type and seed color:

1. **Black and purple eye**: The immature pods are shelled easily as the pliable hull and seeds come out of the pod freely. The shelled peas are attractive, mild flavored and suitable for processing. The white hilum is surrounded by black, pink or light red.
2. **Brown eye:** Pods vary in color from green to lavender and have a wide range of lengths. The immature seeds on cooking turn medium to dark brown in color with delicate flavor.
3. **Crowder:** Seeds are closely crowded in the pods and tend to be globular shaped.
4. **Cream:** Seeds of these types are generally cream colored and have no noticeable "eye" (the hilum is inconspicuous).
5. **Clay:** These are older types with medium to dark brown color and kidney shaped seeds.
6. **White acre:** The peas are kidney shaped with a blunt end. This type is a semi-crowder, generally tan in color and somewhat small. Pods are quite stiff.

The dried purple eye or black eye types are used as food products. The dried beans are frequently sold directly to the consumers on cleaning and bagging. Another common product is the canned product, which is cooked with water prior to canning. Various soups and bean mixes will incorporate this product as well. The cowpea seed protein is rich in the amino acids, particularly, lysine and tryptophan, compared to cereal grains, however, compared to animal protein cowpea proteins are deficient in methionine and cystine contents. Therefore, cowpea seed is valued as a nutritional supplement to cereals and an extender of animal proteins.

In many parts of the world, cowpea is regarded as the only available high quality legume hay for livestock feed. Digestibility and yield of certain cultivars have been shown comparable to alfalfa. It is suitable for grazing and may be grazed during the growing season, while still leaving sufficient material for hay or silage production at the end of season. It may be used as green or dry fodder. The incorporation of cowpea organic manure into the soil, leads to improvement in soil structure, water infiltration rate and soil water holding capacity. Organic matter provided by this legume is quickly decomposed by soil microorganisms and does not persist for longer period in the soil. Resulting organic matter will also increase the ability of the soil to hold plant nutrients by improving the soil cation exchange capacity. It is also used as a green manure, a nitrogen fixing crop or for erosion control. Similar to other grain legumes, cowpea contains trypsin inhibitors, which limit protein utilization.

Cowpea can be harvested at three different stages of maturity: green-snapes, green-mature and dry. Depending on temperature, fresh market (green-mature) peas are ready for harvest 16-17 day after bloom. Harvest date of green-snap pods is normally specified by the processor.

Agronomy

Cowpea thrives better in temperature range of 27^0 to 35^0C and can be grown in acidic to neutral soils but appears less adapted to alkalinity. Optimum sowing time for summer crop is March-April, while for *kharif* crop from middle of June and extends upto the end of July depending on the onset of monsoon. For grain or vegetable purposes, 20-25 kg ha^{-1} seed is required for sowing of pure crop, whereas for fodder and green manuring, the seed rate should be 35-40 kg per hectare. Several cowpea varieties are grown in various parts of country. The promising ones are V-585, V-240, RC-19, FS-68, C-152, RC-101, GCH-4 and GCH-6. The varieties exclusively recommended for fodder purposes are Russian Giant, K-397, K-585, EC-4216, FOS-1, FS-145, S-450 and S-457. Being a leguminous crop, it needs only a small quantity of nitrogen for early growth period on those soils which are poor in organic matter. Application of phosphorus is very important as it promotes proper development of roots and function of *Rhizobium* bacteria.

Being a rainy season crop, no irrigation is required but good drainage is essential. Effective control of weeds in the first 20-25 days of the crop season is required. Green pods as a source of vegetable are harvested at the tender stage otherwise delay may lead to development of fibres. For grain purpose the crop is harvested on maturity of pods or when they turn green-yellow. A good crop of cowpea may yield about 12-15 quintals of grain and 50-60 quintals of straw per hectare. If the crop is raised for fodder purposes, 250-300 quintals of green fodder is obtained from one hectare pure crop.

Abiotic Stresses

Osmotic and passive water uptake abilities of cowpea are generally lower, than other pulses. However, cowpeas have deeper root system and roots further develop vertically, plants are therefore, gradually exposed to water stress, still cowpea has low water uptake ability, hence is suited nicely to water scarce situations. Further, screening of some important crops to drought at seedling stage based on percentage of dead plants, soybean appeared the most susceptible and cowpea the most drought tolerant, as per the following rankings:

Soybean < black gram < moong bean < ground nut < maize < sorghum < pearl millet < bambara nut < lab lab bean < cowpea (Singh *et al.*, 1999).

Cowpea are rarely insensitive to photo period but are typically quantitative short day plants wherein photoperiods longer than 12.5 and 13.5 h d^{-1} increase the number of days from sowing to flowering. Both photoperiod insensitive and photoperiod-sensitive genotypes flower sooner in warmer than cooler temperature (Warrang and Hall, 1983).

Cowpea lines are available that have heat tolerance during early flowering but are susceptible during pod set. Night temperature in the range of 25-30^{0}C induce substantial damage to heat susceptible cultivars during two distinct stages of reproductive development: floral bud development and anther development.

Major diseases and insect-pests

Cowpea crop is affected by a number of fungal, bacterial and viral diseases at different growth stages. The major fungal diseases are: Anthracnose (*Colletotrichum lindemuthianum*), Charcoal rot (*Macrophomina phaseolina*), Sclerotinia rot (*Sclerotium rolfsii*), Fusarium Wilt (*Fusarium oxysporum fsp. Tracheiphilum*), Cercospora leaf spot (*Cercospora canescens*), Brown rust/leaf rust (*Uromyces phaseoli, U. appendiculatus*), Powdery mildew (*Erysiphe polygoni*), Web blight, Seedling rot (*Corticium solani, Rhizoctonia solani*), and Septoria leaf spot (*Septoria vignae, S.vignicola*). Dry; warm and high humid conditions are more conducive for disease development and dissemination of their fungal spores. Among viral diseases: cowpea yellow mosaic, cowpea mild mottle, cowpea cucmo-virus, southern bean mosaic and cowpea tobacovirus are important. A characteristic symptom of the mosaic virus disease is an intermixing of light and dark-brown areas. Infected plants may be dwarf, bushy and giving poor yields. Mosaic diseases can also result in malformed pods. Fungal and viral diseases can be reduced using a rotation with other crops for four to five years and planting of certified seeds of resistant varieties. While bacterial blight is caused by the bacteria *Xanthomonas oxonopodis* pv. Vignicola is the major disease in tropical and subtropical parts of Africa, USA and India.

Almost every part of the cowpea plant is infested by one or another insect species. The important ones causing considerable damage to the crop are white fly, jassids, leaf damaging weevil, pod borers, cutworms, hairy caterpillars, whitegrubs, termites, grasshoppers and bruchids. In Rajasthan, cowpea is exclusively grown in the kharif season from July to October. Population of hairy caterpillar sporadic in appearance is maximum during the second and third week of July. Jassids appear on the crop during second and third week of August and their population generally rises to the peak in a fortnight only. The whitefly acts as a vector for yellow mosaic virus. Weevils attack is more severe, when there is dry spell of at least 25-30 days, while grasshopper are found as minor pests during the entire cropping season. Aphid infestation depends on the humidity, cloudy weather and low sunshine hours during August to September. Bruchids or storage pest are more damaging.

REFERENCES

Davis, D.W., Oelke, E. A., Oplinger, E.S., Doll, T.D., Hanson, C.V. and Putnam, D.H. 1991. *Alternative Field Corps Manual; Cowpea.* Centre for Alternative Plant and Animal Products, Minnesota Extension Service, University of Minnesota, St. Paul, MN 55108.

Fery, R.L. 1990. The cowpea: Production, utilization, and research in the United States. *Hort. Rev.* 12: 197-222.

Hall, A.E. and Patel, P.N. 1987. Cowpea improvement in semi-arid regions of sub-saharan Africa, pp. 279-290. *In: Food grain production in semi-arid Africa* (Eds. J.M. Menyongal) *Proc. Int. Drought Symp. "Food Grain Production in Semi-arid Africa,* Nairobi, Kenya, May, 1986, SAFGRAD, Nairobi.

Marechal, R., Mascherpa, J.M. and Stainer., F. 1978. Etude taxonomique d'un groupe complexed' especesdes generes *Phaseolus* et *Vigna* (Papilionaceae) sur la base de donnees morphologiques et polliniques, traitees par 1' analyse informatique. *Boissiera* 28: 1-273.

Mishra, S.N., Verma, J.S. and Jayasekara, S.J. B. A. 1985. Breeding cowpea to suit Asian cropping systems and consumer traits. pp. 193-193 *In: Cowpea Research, production and utilization* (eds. S.R. Singh and K.O. Rachie) John Wiley and Sons, U.K.

Padulosi, S. and Ng, N.G. 1997. Origin, taxonomy and morphology of *Vigna unguiculata* (L.) Walp. pp. 326-332. *In Advances in cowpea research* (Eds. B.B. Singh, D.R. Mohan Raj, K.E. Dashiell and LEN Jackai), Co-publication of International Institute of Tropical and Japan International Research Centre for Agricultural Sciences, IITA, Ibadan, Nigeria.

Ricardo, B. 1985. Nutritive value of cowpea. *Cowpea Research, Production and Utilization.* (S.R. Singh and K.O. Rachie eds.): John Wiley and Sons Ltd.

Singh, B.B. and Natore, B.R. 1985. Development of Improved cowpea varieties in Africa. pp. 105-115. In: *Cowpea Research, production and utilization* (eds. S.R. Singh and K.O. Rachie). John willey and Sons, U.K.

Singh, B.B., Mai-Kodomi, Y. and Terao, T. 1999. Relative drought tolerance of major rainfed crops of the arid tropics. *Indian J. Genet.* 54 (4): 437-44.

Steele, W.M., Allen, D.J. and Summerfield. R.J. 1985. Cowpea (*Vigna unguiculata* (L.) Walp), pp. 520-583. In: *Grain legume crops*. (R.J. summerfield and E.H. Roberts eds.) William Collins Sons & Co. Ltd., London.

Verdcourt, B.1970. Studies in the *Leguminosae-Papilionoideae* for 'flora of Tropical East Africa': IV. Kew Bul. 24: 507-569.

Warrang, M.O.A. and Hall, A.E. 1983. Reproductive response of cowpea to heat stress: genotype differences in tolerance to heat at flowering. *Crop. Sci.* 23: 1088.

Westphal, E. 1974. Pulses in Ethiopia; their taxonomy and agricultural significance. In: *Agricultural Research Report*. Centre for Agril. Publishing and Documentation, Wageningen, the Netherlands, pp. 213-232.

CHAPTER - 2

COWPEA GENETICS

R.S. Sangwan

SUMMARY

Cowpea is of major importance to the livelihoods of millions of relatively poor people in the developing countries. It is utilized for food, feed and as a cash crop along with spillover benefits to the farmland. It is a cheap source of nutritive value for poor people. The genetics of cowpea have been reviewed by many workers. This chapter covers the work done on origin, history, taxonomy cyto-genetics and genetics in India and abroad of last 50 years. Considerable efforts have been made to review the inheritance studies in cowpea on leaf, flower, maturity, pigmentation, pod and seed traits, grain yield, fodder yield, quality and yield components. Future thrusts and breeding implications for developing ideal plant type in cowpea have been suggested.

INTRODUCTION

Cowpea *(Vigna unguiculata (L.) Walp))* is an important food legume. Being a drought tolerant crop with better growth in warm climates, it is most popular in the semi-arid regions of the tropics, where other food legumes do not perform well. It is a multipurpose crop, for instance, its grains are used as food, fresh pods of some varieties as vegetable and leaves as spinach. It is also utilized for fodder, green manuring and as a quick growing cover crop under wide range of conditions. The cowpea is a good companion crop with a wide range of food, fodder and fibre crops. With potential to fix atmospheric nitrogen upto 240 kg/ha, the cowpea meets a high proportion of its own requirements besides, leaving a fixed N deposit in the soil upto 60-70 kg/ha for succeeding crop (Rachie, 1985).

On a global basis cowpea is one of the important pulse crops. It is estimated that cowpea is cultivated on almost 12.5 million hectares, with an annual production of over 3 million tons worldwide (Singh *et al.*,1997). This food legumes is grown on some 80000 hectares in U.S. However, Central and West Africa account for more than half of the cultivated area, followed by Central and South America, Asia, East and Southern Africa (Singh *et al.*, 1997). In Asia major cowpea producing countries include India, Sri Lanka, Bangladesh, Myanmar, China, Korea, Indonesia, Nepal, Pakistan, Philippines, Thailand and Malaysia with an estimated area almost 1.2 million ha (Mishra *et al.*, 1985). Among the Asian countries, India alone is estimated to cultivate 0.5 million ha for all purposes green pods, dry seed, fodder, green manure and cover crops (Rachie, 1985).

1. Origin and History

Cowpea is one of the ancient human food sources and has probably been used as a crop plant since Neolithic times. The precise location of its center of origin is rather difficult to mention. The previous speculations on the origin and domestication of cowpea have however been, based on botanical and cytological evidences and information on its geographical distribution, cultural practices and historical records (Faris, 1965; Steele and Mehra, 1980; Ng and Marechal, 1985; Ng, 1995). Based on the range of variation and number of varieties found in wild cowpea, as well as primitive characteristics, (perenniality, hairiness, small size of pods and seeds, pod shattering) the highest genetic diversity and most primitive wild *V. unguiculata* occur in southern Africa in the region encompassing Namibia from West, across Botswana, Zambia, Zimbabwe and Mozambique to the east and the Republic of South Africa. Probably, the transversal region of the Republic of South Africa was the center of speciation of *V. unguiculata*, due to presence of most primitive wild varieties, : var. rhomboidea, var. protracta, var. tenuis and var. stenophylla (Padulosi and Ng, 1997). India is the modern center of diversity of cowpea cultivars (Pant *et al.*, 1982; Steele, 1976). Variety *dekindtiana* is believed to be the probable progenitor of the cultivated cowpea (Rawal, 1975; Lush, 1979; Steele and Mehra, 1980; Ng and Marechal, 1985). However, it is not certain to what extent the other wild varieties or subspecies of *V. unguiculata* have contributed to the origin and diversity of cowpea. In the African context, the role of cowpea is predominantly as a pulse, however, after the introduction of *unguiculata* form to India and southeast Asia, two other cultigroups were evolved: *sesquipedalis* (vegetable type) and *biflora* (fodder type) under the influence of human selection (Smartt, 1985). Sporadic occurrence of *sesquipedalis* in India indicates its recent introduction from southeast Asia (Pant *et al.*, 1982). Steele and Mehra (1980) concluded that cultivars grown in India for fodder and seed are integrates of *unguiculata* and *biflora* cultigrops, while those grown for green pod as vegetable are integrates of *unguiculata* and *sesquipedalis*. The *textiles*, a primitive cultivar, was probably selected from *unguiculata* in West Africa for fibre form its long peduncle (Ng and Marechal, 1985).

Cowpea as a food legume, is an important source of nutrients and also provides high quality, less expensive protein to cereal grains or starchy foods based diets. Protein concentration in general, ranges from about 3 to 4 per cent in green leaf; 4-5 per cent in immature pods and 25-30 per cent in dried seeds. Amino acid profile reveals that lysine, leucine and phenylalanine contents area relatively high in cowpea (Bressani, 1985) however, it is deficient in methionine, cystine and tryptophan contents (Boulter *et al.*,1973). Cowpea however contains some antiphysiologic substances, such as tripsin and chymotrypsin inhibitors, cyanogenic glucoside and condensed tannins (Laurena *et al.*, 1984; Montgomery 1964) which decrease protein digestibility and reduce its quality. In tropical Africa and Latin America cowpea is primarily used in the form of dry seed cooked as a pulse in large variety of dishes and in Asian countries it is cultivated both for green pods for vegetable uses and dry seed as pulse.

2. Taxonomy

Cowpea belongs to the botanical species *Vigna unguiculata* L. Walp. under the subfamily papilionoideae of family Leguminosae with 2n=22. There is considerable confusion in the synonymy and names of the cultivated *Vigna* species. Verdcourt (1970) concluded that cowpea (formerly *V. sinensis* (L.) Saviex Haask.), catjang (formerly *V. cylindrical* (L.) Skeels), and asparagus bean (formerly *V. sesquipedalis* (L.) Fruw.) are not to be regarded as distinct species since they cross freely to from fully fertile hybrids. It was suggested that these taxa should be treated as subspecies of *V. unguiculata* (L.) Walp., i.e. cowpea = *V. unguiculata* subsp. *Unguiculata,* catjang = *V. unguiculata subsp.,* Catjang, and asparagus bean = *V. unguiculata subsp., sesquipedalis.* Marechal *et al.,* (1978) subdivided the genus described earlier by Verdcourt (1970) into seven subgenera.

In this classification, *V. unguiculata* (L.) Walpers and *V. nervosa* Markotter constitute the section catjang, one of the six sections of subspecies *Vigna* and species of this section are characterized by spurred stipules below the attachment point of the leaf stalks and can be shaped peel with beak. The surface of their pollen grains are reticulate with raised exine (DeLeonardis *et al.,* 1993). Interspecific crosses between two species have not been successful (Mithen, 1987). Thus, all the cultivated cowpeas are grouped under *V. unguiculata* subspecies *unguiculata.* Reproductive barriers between these subspecies are weak (Rawal *et al.,* 1975; Sakupwanya *et al.,* 1989), while they are strong between *V. unguiculata* and closely related species like *Vigna vexillata* (Barone *et al.,*1992). Cultivated cowpea originated in Africa, the only continent where wild forms are encountered (Faris, 1965; Marechal *et al.,* 1978). Wild cowpea is genetically highly variable across Africa. Although morphological characters overlap somewhat, they are sufficiently discrete leading to proposed formal taxa. Morphological studies (Pienaar and Van Wyk, 1992; Mithen and Kibblewhite, 1993; Pasquet, 1993a) show some divergences, but opine that cowpea gene pool consists of several perennial taxa and one annual taxon. The last taxonomic treatment (Pasquet, 1993a,b, 1997) divides cowpea into ten perennial subspecies and one annual subspecies, ssp.

unguiculata. It splits ssp. *Unguiculata* into var. *unguiculata* (cultivated cowpea) var. *spontanea* (Schweinf.) and *pasquet* (annual wild cowpea). The classification and nomenclature of the wild taxa *V. unguiculata*, is however, complicated and more than 20 epithet names have been used in the past (Padulosi and Ng, 1997).

According to their breeding systems (Pasquiet, 1999), these 11 subspecies of *unguiculata* have been classified into 3 groups : the perennial outcross, the perennial out-inbreds and the inbred annuals; the allozyme data confirm this grouping in 199 germplasm accessions of wild and cultivated cowpea.

3. Botany

Cowpea is an annual herb with a strong tap root system with different growth habits i.e. erect, semi erect, trailing or climbing, bushy annual with glabrous stem. The first pair of leaves above the cotyledonary node is simple and opposite, they however exhibit considerable variation in size and shape (Faris, 1965). The trifoliate leaves arise alternately and terminal leaflet is frequently longer and of greater area than that of asymmetrical lateral leaflets. There is a wide range of leaf size and shape from lanceolate to ovate. Leaf orientation is usually planophile in cultivated species, becoming more erectophile in the wild types.

Flowers are racemose or indeterminate inflorescence, at the distal ends of peduncles which are in leaf axil. Terminal inflorescences are rare, perhaps only occurring in subsp. *cylindrical*. The flower is typical with five sepals in gamosepalous condition and five petal in polypetelous condition of (1+2+2). The outermost posterior petal is standard, two laterál petals (wings) are clawed almost covering two interior petals which are loosely connate to form boat shaped keel.

Both androecium and gynaecium are enclosed inside the keel petals. The gynaecium is monocarpillary. Ovary is sessile, multilocular and grows out into a pod after fertilization. The style is longitudinally beared along inner side and ends in oblique stigma and androecium is diadelphous consisting of 10 (9+1) stamens. The flowers are borne in alternate pairs and although 6-12 pairs may occur per inflorescence frequently only the first two pairs develop. The flowers are borne on short pedicels and corollas vary in colour from to purple, the former being mainly associated with white or partly white seeds. Flowers generally open before dawn (4.30-5.00 a.m.) and most of them shut by noon, with sporadic flower opening in the afternoon (Hazra *et al.*, 1990). The dehiscence of anthers is much earlier to anthesis. Krishaswamy *et al.* (1945) observed variable time of dehiscence from 10.00 PM to 0.45 AM and this variation may be affected by environmental factors like moonlight, clear sky and dry warm atmosphere which result self pollination in cowpea.

Seeds vary considerably in size, shape and colour, generally they are 2-12 mm long and weigh 5-30 g/100/seeds. Seeds shape is correlated with that of pods where, individual seeds are adjacent ones during development, they become kidney shaped, but as crowding within the pod increases the seeds become globular and are called crowders. The testa may be smooth or wrinkled with white, buff, green, red, brown, black, speckled, mottled, blotched colours.

4. Cytogenetics and Cytology

Vigna unguiculata is a diploid species, with 2n = 22. Within the genus *Vigna, V. unguiculata* is a homogeneous species, characterized by striking morphological features like a spurred stipule, an almost symmetric purple flowers (sometimes white in cultivated forms) a style with short beak and a curved pod beak (Marechal *et al.*, 1978). Meiotic chromosomes at pachytene stage were classified 11 bivalents into one short (19 µm), seven medium (26-34 µm) and three long (41-45 µm) chromosome and the chromosomes were not distributed uniformly along the chromosome arms (Mukherjee, 1968). Induced tetraploidy studies revealed that meiotic abnormalities in autopolyploids such as quadrivalent, trivalent and univalent formations and occurrence of laggards and unequal distribution in the anaphase and resulting impaired fertility and concluded that it is highly improbable to produce economically useful tetraploids in cowpea (Sen and Bhowal, 1960).

Crosses of *Vigna unguiculata* with other species of *Vigna* have largely been unsuccessful (Sen and Bhowal, 1960; Faris,

1965). Crosses between *mensensis* and domesticate subsp resultant seeds failed when the former was the female parent and while using the former as male shriveled seeds and seedlings of low vigour were produced (Rawal *et al.*,1975). Premsekar *et al.* (1964) while attempting reciprocals between *V. sesquipedalis* and *V. sinensis* observed hybrid resembling with *V. sinensis* which had 42 per cent pollen sterility despite normal meiosis. Mahudeswaran *et al.* (1973) derived 'CO-2' variety from a three way crosses involving *V. catjang, V. unguiculata* and *V. sesquipedalis* which was high yielding and quite fertile. No viable interspecific hybrids have been reported among other species of *Vigna* group (Singh *et al.*, 1964) such as *V. umbellate, V. vexillata, V. mungo, V. radiata, V. acontifolia* and *V. angularis*. The intergeneric crosses between *V. unguiculata* and *Phaseolus cocineus* were not successful (Smartt, 1990).

Cowpea karyotype has been studies in detail by Barone and Saccardo, (1990), Pignone *et al.*,(1990) and Saccardo *et al.*, (1992). Barone and Saccardo (1990) used pachytene bivalent to develop their karyotype, while Pignone *et al.* (1990) developed a banded karyotype by using cells in mitotic prometaphase. Saccardo *et al.* (1992) used both conventional techniques and automatic image analyzer system as well, to develop karyotype with pachytene and mitotic prometaphase and metaphase chromosomes.

Fatokun and Singh (1987) successfully crossed cultivated cowpea with a hairy wild relative (*Vigna pubescence)* using tissue culture technique to reuse the hybrid embryos which would otherwise have been shriveled and degenerated. Resultant F1 plants were vigorous in growth but partially sterile and showed some meiotic abnormalities like few univalents and quadrivalents in the pollen mother cells. Barone and Ng (1990) unsuccessfully attempted interspecific cross between *V. unguiculata and V. vexillata* for which important barriers were lack of fertilization and collapse of fertilized ovules 5-8 days of pollination. Similarly attempts made by Fatokun (1991) to cross of *V. vexillata* with various cultivated and non-cultivated cowpeas turned unsuccessful.

In recent years, considerable progress has been made in developing innovative biotechnologies for cowpea. The focus of this research is on understanding of cowpea genome and development of techniques to insert exotic genes into cultivated cowpea (Monti *et al.*, 1997; Fatokun *et al.*, 1997.

In interspecific crosses made between *V. vexillata* and *V. unguiculata;* the cytological studies of the meiotic chromosomes revealed high frequency of univalent formation in F1 which indicated that genomes of paretal species were structurally different and hybrid also exhibited enhanced enzyme activities of peroxidase, polyphenol oxidase and phenylalanine ammonia lypase (Gomathinayagam *et al.*, 1998).

5. Genetics

Cowpea genetics has been reviewed by Fery (1985), Sharma and Joshi (1993) and Fery and Singh (1997) on cytology, qualitative and quantitative traits. A review on the on cowpea research are presented below genetics in India during last fifty years on different aspects.

(i) Heterosis

Heterosis breeding is an important genetic tool for exploiting the predominantly non-additive gene action. In self-pollinated crops, utilization of hybrid vigour is dependent upon a system for economically producing F1 seeds. Hybrid vigour in cowpea, as in other crops, depends on specific presents used in hybrid combination. Several studies on cowpea have shown considerable heterosis for many characters. For instance, heterosis was demonstrated for leaves per plant, dry and fresh stem weight and dry and fresh fodder yield (Singh *et al.*, 1986), days to flowers, vine length, stem girth, leaf length, leaf breadth, green fodder yield, dry matter yield (Lodhi *et al.*, 1990), branches per plant and plant height (Sawant *et al.*,1994a), number of clusters per plant (Patil and Shete, 1987; Viswanatha *et al.*,1998), number of pods (Bhaskaraiah *et al.*, 1980, Zaveri *et al.*,1983; Sawant *et al.*,1994; Viswanatha *et al.*,1998), number of seeds per pod (Patil and Shete, 1987; Sangwan and Lodhi, 1995; Viswanatha *et al.*,1998), seed weight (Patil and Shete, 1987; Sangwan and Lodhi, 1995; Viswanatha *et al.*,1998), grain yield (Bhaskaraiah *et al.*,1980; Zaveri *et al.*,1983; Patil and Shete, 1987; Sawant *et al.*,1994; Sangwan and Lodhi, 1995; Bhor *et al.*,1997; Viswanatha *et al.*,1998), protein content and *in vitro* dry matter digestibility of fodder (Jain *et al.*,1980; Lodhi *et al.*,1990).

(ii) Variability

Genetic variability is of paramount importance in crop plants for its use, conservation and for bringing continuous improvement in crop plants (Simmonds, 1962). India being center of diversity, rich variability has been observed by various workers for many traits in cowpea viz., plant height (Patel, 1973; Veeraswamy *et al.*,1973; Angadi *et al.*, 1979; Gupta and Lodhi, 1979; Pandita *et al.*,1982; Roquib and Patnaik, 1990; Thaware *et al.*, 1992; Chattopadhyay *et al.*, 1996; Singh *et al.*, 1997); days to flower (Singh and Mehndiratta, 1969; Hanchinal *et al.*, 1981; Pandita *et al.*, 1982; Roquib and Patnaik, 1990; Rewale *et al.*, 1995; Chattopadhyay *et al.*, 1996; Singh *et al.*, 1997; Hazra *et al.*, 1999); days to maturity (Singh and Mehndiratta, 1969; Patil and Baviskar, 1987; Radhakrishnan and Jebaraj, 1992; Rewale *et al.*, 1995; Chattopadhyay *et al.*, 1996; Gunasekaran *et al.*, 1998); number of branches per plant (Chattopadhyay *et al.*, 1996; Vardhan *et al.*, 1998; Trehan *et al.*, 1970); pods per plant (Chattopadhyay *et al.*, 1996; Singh *et al.*, 1997); number of seeds per pod (Kandaswamy *et al.*, 1989; Roquib and Patnaik, 1990; Radhadrishan and Jebaraj, 1992; Rewale *et al.*, 1995; Chattopadhyay, 1996); Pod length (Singh *et al.*, 1997; Hazara *et al.*, 1999); 100- seed weight (Rewale *et al.*, 1995; Chattopadhyay *et al.*, 1996; Singh *et al.*,1997); grain yield per plant (Sawant, 1994; Mathur, 1995; Rewale *et al.*, 1995; Chattopadhyay *et al.*, 1996; Singh *et al.*, 1997; Vardhan and Savithramma, 1998; Gunasekaran *et al.*, 1998) and leaf number (Gupta and Lodhi, 1979).

(iii) Heritability

Heritability is of interest primarily as a measure of the value of selection for particular character in various types of progenies and as an index of transmitability. Heritability in broad sense i.e. the ratio of the total genotypic variance to the total phenotypic variance provides a good parameter for reliable estimate of genetic potential of a given population. Medium to high (17.03 to 99.75%) heritability estimates have been observed for grain yield per plant in different studies but in most of the cases heritability estimates were high. Days to flower and days to maturity in general have high estimates (58.10 to 98.54%) of heritability. Plant height had moderate to

Table 1. Estimates of broad-sense heritability (%) for various characters in cowpea.

Characteristics									
Days to flower	**Days to maturity**	**Plant height**	**Branches/ plant**	**Pods/ plant**	**Seeds/ pod**	**Pod. Length**	**100-seed weight**	**Yield/ plant**	**Reference**
-	-	-	-	-	23.50	24.60	-	-	Trehan *et al.* (1970)
-	-	-	-	-	74.30	94.80	-	-	Sahoo *et al.* (1971)
-	-	-	-	-	33.50	97.50	-	-	Veeraswamy *et al.* (1973)
-	-	-	-	-	78.90	-	-	-	Bapna and Joshi (1973)
-	-	-	-	-	50.60	59.80	-	-	Bordia *et al.* (1973)
-	-	-	-	-	70.10	95.40	-	-	Laxmi and Goud (1977)
-	-	-	-	-	86.30	95.00	-	-	Tikka *et al.* (1977)
69.20	-	-	-	-	40.30	95.00	-	-	Sree Kumar *et al.* (1979)
98.54	-	74.65	71.76	74.22	95.05	99.89	93.49	87.41	Hanchinal *et al.* (1981)
-	95.60	15.70	-	-	-	32.20	-	-	Pandita *et al.* (1982)
90.00	-	-	68.80	96.80	78.60	85.50	97.90	-	Jana *et al.* (1982)
88.70	-	-	-	20.70	50.40	65.70	81.90	27.10	Kumar and Mishra (1983)
90.00	-	78.0	-	64.00	-	-	-	57.00	Dumbre *et al.* (1983)

-	-	53.80	-	53.01	64.00	87.37	85.38	35.52	Dharmalingam & Kadambavana-sundaram (1984)
58.10	65.00	27.35	34.94	13.97	65.34	62.47	82.47	17.03	Apte *et al.* (1987)
65.30	82.90	-	-	68.70	-	70.30	90.90	51.60	Patil and Baviskar (1987)
98.00	-	-	-	-	-	-	-	-	Sharma *et al.* (1988)
-	-	97.60	96.16	98.62	99.60	98.01	-	99.75	Thiyangarajan *et al.* (1989)
91.00	60.00	94.00	-	-	-	76.00	22.00	74.00	Roquib and Patnaik (1990)
97.75	95.48	92.33	-	92.75	75.31	84.06	94.43	94.66	Siddique and Gupta (1991b)
-	-	-	-	80.30	70.20	86.30	-	85.90	Gowda *et al.* (1991)
-	87.80	97.90	96.50	98.90	94.50	91.00	99.60	-	Radhakrishnan & Jebaraj (1992)
95.80	-	90.60	-	-	-	-	-	-	Sharma and Singhanai (1992)
98.40	96.30	97.00	86.00	97.50	69.10	78.40	99.20	98.50	Selvi *et al.* (1994)
75.35	87.60	58.25	31.91	36.52	51.15	25.22	82.60	28.50	Rewale *et al.* (1995)
-	-	99.86	-	98.10	94.02	94.92	-	97.30	Mathur (1995)
92.36	92.19	91.73	-	93.94	72.83	83.79	95.76	75.22	Singh *et al.* (1997)
87.33	89.69	97.60	69.94	72.83	83.79	95.76	96.76	75.22	Kumar (1998)
96.24	80.61	96.05	-	-	75.00	-	-	81.42	Sharma (1999)

high heritability estimates in most of the studies. Moderate to high heritability accompanied by high genetic advance was reported for number of branches per plant. The numbers of pods per plant had low to high (13.97 to 98.90%) heritability estimates, wheras seeds per pod had moderate to high heritability in concominant with moderate genetic advance. 100-seed weight had high heritability estimates with a high genetic advance (Table 1).

(iv) Inheritance

The information on the inheritance of important plant traits would beneth way for their genetic important. Study on the inheritance of different morphological, grain, fodder and quality traits have been conducted by various workers which have been summarized Table 2.

Table 2. Inheritance of important traits in cowpea

Character	Genetic architecture	Reference
Growth habit /vine length	Two duplicate genes control indeterminate growth habit	Kolhe (1970).
	Trailing growth habit is conditioned T_1, T_2 and T_3 genes where T_1 and T_2 are complementary	Singh and Jindla (1971).
	Viny type dominant over bushy plant type	Singh *et al.*, (1976).
	Monogenic inheritance for growth habit	Premsekar & Raman (1972).
	Predominance of non-additive genetic variance	Jain *et al.*, (1981), Sawant (1994 b), Ponmariamma & Das (1997).
	Sizeable amount of both additive and non-additive genetic variance	Jhorar *et al.*, (1990), Sangwan *et al.*, (1998), Bhor & Dumbre (1998).
Branch number	Additive genetic variance	Jain *et al.*, (1981).
	Non-additive genetic variance	Sawant (1994b), Ponmariamma and Das (1997).
	Additive and non-additive	Jhorar *et al.*, (1990), Sangwan *et al.*, (1998).

Leaf characters	Large leaf dominant over small leaf	Krishnaswamy *et al.*, (1945).
	Long leaf dominant over short leaf	Kohle (1970).
	Hastate dominant over globose and controlled by 4 genes	Jindla and Singh (1970).
	Non-additive genetic variance for leaflet length and breadth	Jain *et al.*, (1981).
	Additive and non-additive genetic variance	Jhorar *et al.*, (1990), Ponmariamma & Das (1997).
	Additive	Sangwan *et al.*, (1998).
Flower colour	Monogenic inheritance	Kohle (1970).
	Violet dominant over light violet flower colour	Jindla and Singh (1970), Venugopal and Goud (1977), Hanchinal and Goud (1978).
	4 genes control flower colour	Biradar *et al.*, (1997).
	Purple dominant over white	Sangwan and Lodhi (1998b).
Days to flower	Lateness dominant over earliness and controlled by two complementary genes	Roy & Richharia (1948).
	Additive gene action	Lal *et al.*, (1975), Tikka *et al.*, (1976), Zaveri *et al.*, (1980), Zaveri *et al..*, (1983), Patil and Bhopkar (1986), Mishra *et al.*, (1987), Sawant (1994 b), Bhor and Dumbre (1998).
	Non-additive gene action	Singh and Dabas (1986), Sangwan (1996), Ponmariamma & Das (1997).
Days to maturity	Additive and non-additive genetic variance	Singh and Dabas (1986).
	Additive	Zaveri (1980), Patil and Bhapkar (1986) and Bhor & Dumbre (1998).
Pod colour	Monogenic brownish straw dominant over straw colour	Sen and Bhawal (1961).
	Green pods dominant over light green pods	Premsekar and Raman (1972)
	Two genes for green pod colour	Hanchinal & Goud (1978

	Three genes	Lokaprakash *et al.*, (1983
	Purple dominant over green and 4 genes control it.	Biradar *et al.*, (1997).
	Black pods partially dominant over white pod colour with monogenic inheritance	Sangwan and Lodhi (1998b).
	Light green dominant over green, pigmantation dominant over no pigmantation and 1 to 5 genes involved in its inheritance.	Venugopal (1998).
Calyx pigmentation	3 duplicate genes	Venugopal and Goud (1977).
	3 genes and purple dominant over green	Biradar *et al.*, (1997).
Pod number	Non-additive genetic variance	Zaveri *et al.*, (1983), Singh and Dabas (1986), Sawant (1994b) and Sangwan (1996).
	Additive	Patil and Patil (1987).
Pod length	Additive and non-additive	Singh and Jain (1972).
	Partial dominance of long pods	Roy & Richharia (1948), Jindla and Singh (1970).
	Partial dominance of narrow pods, 6 gene control pod length	Bhowal (1976).
	Multiple gene control	Singh *et al.*, (1976)
	Additive genetic control	Patil and Patil (1987), Sawant (1994b) and Sangwan (1996).
	Non-additive genetic variance	Roy & Richharia (1948), Goswami and Kumari (1972), Bhor & Dumbre (1998).
Seeds per pod	Additive and non-additive genetic variance	Patil and Patil (1987), Sangwan (1996).
	Non-additive genetic variance	Sawant (1994b), Bhor and Dumbre (1998).
Seed colour	Black dominant over red	Roy and Richharia (1948), Kolhe (1970).
	One recessive gene control	Singh *et al.*, (1976).

	Digenic inheritance	Premsekar and Raman (1972), Sangwan *et al.*, (2000).
Seed weight (size)	Partial dominance of small seeds 4 gene control for seed size	Bhowal (1976).
	Predominance of additive gene Bold seeds dominant over small sized seeds	Jatasra (1980), Singh & Dabas (1986), Sawant (1994b), Sangwan and Lodhi (1999).
	Additive and non-additive	Patil and Patil (1987).
	Large dominant over small seeds with monogenic control	Karkannavor *et al*,, (1991).
	Predominantly non-additive	Sangwan&Lodhi (1998a)
Grain yield	Predominance of non-additive genetic variance	Zaveri *et al.*, (1983), Singh and Dabas (1986), Mishra *et al.*, (1987), Sangwan (1996), Bhor and Dumrbe (1998).
Fodder yield	Additive and non-additive with predominance non-additive gene effects for green fodder and dry matter yields	Jatasra (1979), Jain *et al.*, (1981), Sangwan *et al.*, (1998).
	Additive and non-additive	Jhorar *et al.*, (1990), Pon mariamma & Das (1997).
Protein	Non-additive genetic variance predominant	Jain *et al.*, (1980), Sangwan *et al.*, (1997), Ponmariamma and Das (1997).
IVDMD	Non-additive genetic variance predominant	Jain *et al.*, (1980), Sangwan *et al.*, (1997).

FUTURE THRUSTS AND BREEDING IMPLICAITONS

An ideal cowpea variety for grain would be preferred having high grain yield, better grain characters and its quality, early maturity, suitability for various cropping systems, disease resistance; whereas for fodder, emphasis should be given on dual purpose varieties having high fodder yield, early maturity, along with reasonably good grain yield with disease resistance. In cowpea, different plant ideotypes are, therefore, required for attaining higher yield as per their use for grain, fodder or vegetable purposes. Therefore, there is need to develop suitable plant types which may give higher yield of grain, fodder or both and a good cooking quality under a particular cropping system. In cowpea, sufficient genetic variability exists for different traits viz., days to flower, days to maturity, plant height, branches per plant, pods per plant, seeds per pod, pod length, seed weight and grain yield per plant. Likewise, there is lot of variability for fodder yield and green pod traits in this crop. Most of the characters except the grain yield and fodder yield have moderate to high heritability and, hence, these could be improved through simple selection.

Higher extent of heterosis for grain yield, number of clusters, pod length, number of pods, seeds per pod, seed weight and a number of fodder traits including fodder yield have been observed. Majority of these traits had strong positive correlation with seed yield and among themselves. Therefore, selection based on these traits will lead to yield enhancement.

Both grain and fodder yields are governed predominantly by non-additive genetic variances. However, additive genetic variance is also present in sizeable quantity and is predominant for pod length, leaf size, days to flower, days to maturity and seed weight. In order to exploit such, both types of genetic variances, some kind of intermating of selects in early generations would be most useful and to break undesirable linkages so as to get desirable recombinants. Moreover, this approach will maintain sufficient variability in the population for its exploitation in later generations and to have better gain through selection.

REFERENCES

Angadi S P, Subramani A and Kulkarni R S. 1979. Multibariate analysis on cowpea (Vingna unguiculata). *Genetic Agrararia* 33:121-132

Apte U B, Chavan S A and Jadhav B B. 1987. Genetic variability and heritability in cowpea. *Indian Journal of Agricultural Sciences* 57: 596-598

Bapna C S and Joshi S N. 1973. A study on variability following hybridization in Vigna sinensis (L.) Savi. *Madras Agricultural Journal* 60: 1369-1372

Barone A and Ng N Q. 1990. Embryological study of crosses between Vigna unguiculata and V. vexillata. 151-160. In Cowpea genetic resources, (edition N.Q. Ng and L.M. Monti), IITA, Ibadan, Nigeria

Barone A and Saccardo F. 1990. Pachytene morphology of cowpea chromosomes. 137-143. In *Cowpea genetic resources* (edition N.Q. Ng and L.M. Monti), IITA, Ibadan, Nigeria

Barone A, Del Giudica A and Ng N Q. 1992. Barriers to interspecific hybridization between Vigna unguiculata and Vigna vexillata. *Sex Plant Reproduction* 5: 195-200

Bhaskaraiah K B, Shiv Shankar G and Virupakashappa K. 1980. Hybrid vigour in cowpea. *Journal of Genetics and Plant Breeding* 40: 334-337

Bhor T J and Dumbre A D. 1998. Gene action of some characters of cowpea. *Legume Research* 21: 177-182

Bhor T J, Kute N S, Dumbre A D and Sarode N D. 1997. Heterosis and inbreeding depression in cowpea. *Indian Journal of Agricultural Research* 31: 122-126

Bhowal J G. 1976. Inheritnce of pod length, pod breadth and seed size in cross between cowpea and catjang bean. *Libyan Journal of Science* 6: 17-21

Biradar B D, Goud J V and Patil S S. 1997. Differential expression of pleiotropic genes for pigmentation in cowpea (*Vigna unguiculata* L. Walp.). *Crop Research* (Hisar) 14(2): 233-242

Borah H K and Khan A K F. 1999. Correlation and path analysis in fodder cowpea. *Crop Research* (Hisar) 18: 278-282

Bordia P C, Yadavendra J P and Kumar S. 1973. Genetic variability and correlation studies in cowpea. *Rajasthan Journal of Agricultural Sciences* 4: 39-44

Boulter D, Evans I M and Thompson A. 1973. Protein quality. Proc. Int. Inst. Tropical Agric. Grain Legumes Imp. Workshop, 29 Oct.-2 Nov., 1973

Bressani Ricardo 1985. Nutritive value of cowpea, In. *Cowpea Research, Production and Utilization,* (Eds. Singh, S R and Rachie, K O), John Wiley and Sons, New York. Pp. 353

Chattopadhyay A, Dasgupta T, Som M G and Hazra P. 1996. Variability studies in vegetable cowpea. *Horticultural Journal* 9:71-75

DeLeonardis Fichera G, Padulosi S and Zizza A. 1993. Preliminary studies on pollen and seed of wild germplasm accessions of Vigna unguiculata (L.) Walpers. *Proceedings 58th Congress of the Italian Botanic Society*, University of Tor Vergata, Rome, 4-8 Oct., Vol. 127:3

Dharmalingam V and Kadambavansundaram M. 1984. Genetic variability in cowpea (*Vigna unguiculata* (L.) Walp). *Madras Agricultural Journal* 71: 640-643

Dumbre A D, Deshmukh R B and Pandhye A P. 1983. Variability in cowpea. *Journal of Maharashtra Agricultural University* 8: 85

Faris D G. 1965. The origin and evolution of the cultivated forms of Vigna sinesis. *Canadian Journal Cytology* 7: 433-452

Fatokun C A. 1991. Wide hybridization in cowpea: Problems and prospects. *Euphytica* 54: 137-140

Fatokun C A and Singh B B. 1987. Interspecific hybridization between Vigna pubescens and *Vigna unguiculata* (L.) Walp. through embryo rescue. *Plant Cell, Tissue and Organculture* 9: 229-233

Fatokun C A, Young N D and Myers G O. 1997. Molecular markers and genome mapping in cowpea. 352-360. In: *Advances in cowpea research*, (Edit. BB Singh, DR Mohan Raj, KE Dashiell and LEN Jackai.) Co-publication of International Institute of Tropical Agriculture and Japan International Research Centre for Agricultural Sciences, IITA, Ibadan, Nigeria

Fery R L. 1985. The genetics of cowpea: a review of world literature. 25-62. In: *Cowpea research, production and utilization* (Eds. S.R Singh and K O Rachie), John Wiley and Sons, Chichester, UK.

Fery R L and Singh B B. 1997. Cowpea genetics: a review of recent literature. 13-29 In. *Advances in cowpea research*, (Eds. BB Singh, DR Mohan Raj, KE Dashiell and LEN Jackai.) Co-publication of International Institute of Tropical Agriculture and Japan International Research Centre for Agricultural Sciences, IITA, Ibadan, Nigeria

Gomathinayagam P, Ganesh Ram S, Rathnaswamy R and Ramaswamy N M. 1998. Interspecific hybridization between *Vigna unguiculata* (L.) Walp. and *V. vexillata* (L.) A. Rich through in vitro embryo culture, *Euphytica* 102: 203-209

Goswami L C and Kumari S. 1972. Studies on breeding behabiour of cowpea (*V. sinensis*). *Journal of Assam Scientific Society* 15: 20-28

Gowda T H, Hiremath S R and Salimath P M. 1991. Estimation of genetic parameters in intervarietal crosses of cowpea (*Vigna unguiculata* (L.) Walp.) and their implication in selection. *Legume Research* 14: 15-19

Gunasekaran M, Selvaraj U and Raveendran T S. 1998. Induced polygenic mutations in cowpea (*Vigna unguiculata* (L.) Walp). *South Indian Horticulture* 46: 13-17

Gupta S P and Lodhi G P. 1979. Variability for fodder yield and its component in cowpea. *Indian Journal of Agricultural Sciences* 46: 407-410

Hanchinal R R and Goud J V. 1978. Inheritance in Vigna. *Indian Journal of Genetics and Plant Breeding* 38: 339-342

Hanchinal R R, Habib A F and Goud J V. 1981. Variability in cowpea (*Vigna unguiculata* (L.). Walp.). *Mysore Journal of Agricultural Sciences* 15: 371-375

Hazara P, Bhattacharya C, Mandal A R and Ghosal K K. 1990. Studies on floral biology of cultivated cowpea. *Indian Biology* 22: 8

Hazara P, Chattopadhyay A and Pandit M K. 1999. Genetic variability in three cultigroup of cowpea. *Journal of Interacademicia* 3: 263-268

Jain S, Lodhi G P and Boora K S. 1980. Combining ability analysis for quality traits in forage cowpea. *Legume Research* 3: 107-111

Jain Subash, Lodhi G P and Boora K S. 1981. Combining ability analysis for forage characters in cowpea. *Forage Research* 7: 131-137

Jana S, Som M G and Das N D. 1982. Genetic variability and correlation studies in cowpea. *Vegetable Sciences* 9: 96

Jatasra D S. 1979. Combining ability analysis for green and dry matter yield in cowpea (*Vigna unguiculata* (L.). Walp.). *Forage Research* 5: 165-168

Jatasra D S. 1980. Combining ability for grain weight in cowpea. *Indian Journal of Genetics and Plant Breeding* 40: 330-333

Jhorar B S, Jatasra D S, Sangwan R S and Lodhi G P. 1990. Combining ability analysis in forage cowpea. *Farm Science Journal* 5: 86-91

Jindla L N and Singh K B. 1970. Inheritance of flower colour, leaf shape and pod length in cowpea (*Vigna sinensis* L.). *Indian Journal of Heredity* 2: 45-49

Kandaswamy G, Kandabavanasundram M and Rajasekaeran S. 1989. Variability in cowpea under different environmental conditions. *Madras Agricultural Journal* 76: 197-199

Karkannavar J C, Venugopal R and Goud J V. 1991. Inheritance of some linkage studies in cowpea (*Vigna unguiculata* (L.). Walp.). *Indian Journal of Genetics and Plant Breeding* 51: 203-207

Kolhe A K. 1970. Genetic studies in Vigna species. Poona Agriculture College Magazine 59: 126-137

Krishnaswamy N, Nambiar K K and Mariakulandai A. 1945. Studies on cowpea [*Vigna unguiculata* (L.) Walp.]. *Madras Agricultural Journal* 33: 145-160

Kumar A and Mishra S N. 1983. Genetic variability for seed yield and other traits in cowpea. *Legume Research* 6: 83-85

Kumar, Ramesh 1998. Genetic variability and correlation studies for grain yield and nutritional attributes in cowpea (*Vigna unguiculata* (L.) Walp.) M.Sc. Thesis, CCS Haryana Agricultural University, Hisar, India

Lakshmi P V and Goud J V. 1977. Variability in cowpea (*V. sinensis*). *Mysore Journal of Agricultural Sciences* 11: 144-147

Lal S, Singh M and Pathak M M. 1975. Combining ability in cowpea. *Indian Journal of Genetics and Plant Breeding* 35: 375-378

Laurena A C, Garcia V V and Mendoza E M T. 1984. Localization and determination of condensed tannins in cowpea. *Philippine Agriculture* 67: 329

Lodhi G P, Boora K S, Jain S and Balchand 1990. Heterosis for fodder yield and quality characters in cowpea *Vigna unguiculata* (L.) Walp. *Crop Research* (Hisar) 3: 66-73

Lokaprakash R, Hiremath S R and Shiv Shankar G. 1983. Differential expression on pleiotropic genes for pigmentation in cowpea. *Mysore Journal of Agricultural Sciences* 17: 337-341

Lush W M. 1979. Floral morphology of wild and cultivated cowpea. *Economic Botany* 33: 442-447

Mahudeswaran K, Veeraswamy R, Palaniswamy G A, Rathnaswamy, R, Rajasekaran, VPA and Marappan, PV. 1973. CO-2 a vegetable cum grain cowpea. *Madras Agricultural Journal* 60: 1844-1845

Marangappanavar L R. 1986. Genetic diversity, gene action and character association in cowpea (*Vigna unguiculata* (L.) Walp.) *Mysore Journal of Agricultural Sciences* 20: 231

Marechal R, Mascherpa J M and Stainier F. 1978. Etude taxonomique d'un grope d' especes des generes Phaseolus et Vigna (Papilionaceae)sur la base donnees morphologiques et polliniques traitees pour 1 analyse informatique. *Boissiera* 28: 1-273

Mathur R. 1995. Genetic variability and correlation studies in segregating generations of cowpea. *Madras Agricultural Journal* 82: 150-152

Mishra S N, Rastogi R and Verma J S. 1985. Note on parental selectivity – a factor affecting success in cowpea crossing. *Trop Legumes Bulletin* 31: 24-26

Mishra S N, Verma J S and Rastogi R. 1987. Combining ability for flowering and seed yield in cowpea. *Annals of Agricultural Research* 8: 268-272

Mithen R and Kibblewhite H. 1993. Tazonomy and ecology of Vigna unguiculata (Lguminosae-papilionoideae) in South-Central Africa. *Kirkia* 14: 100-113

Mithen R F. 1987. The African gene pool of Vigna 1 V. nervosa and V. unguiculata from Zimbabwe. *FAO/IBPGR Plant Genetic Resources Newsletter* 70: 13-19

Montgomery R D. 1964. Observations of the cyanide content and toxicity of tropical pulses. *West Indian Medical Journal* 13: 1-11

Monti L M, Murdock L L and Thottappilly G. 1997. Opportunities for biotechnology in cowpea. Pages 341-35. In: *Advances in cowpea research* (Eds. BB Singh, DR Mohan Raj, KE Dashiell and LEN Jackai). Co-publication of International Institute of Tropical Agriculture and Japan International Research Centre for Agricultural Sciences, IITA, Ibadan, Nigeria

Mukherjee P. 1968. Pachytene analysis in Vigna, chromosome morphology in Vigna sinensis. Science and Culture 34: 252-253

Ng N G. and Marichal R. 1985. Cowpea taxonomy, origin and germplasm. Pages 11-21. In: *Cowpea research, production and utilization* (Eds. S.R. Singh and K.O. Rachie). John Wiley and Sons, Chichester, U.K

Ng N G. 1995. Cowpea *Vigna unguiculata* (Leguminosae Papilionoideae) Pages 326-332. In: *Evolution of crop plants, 2nd edition* (Eds. J. Smartt and N.W. Simmonds), Longman Harlow, U.K.

Padulosi S and Ng N G. 1997. Origin, taxonomy and morphology of *Vigna unguiculata* (L.) Walp. Pages 1-12. In: *Advances in cowpea research.* (Eds. BB Singh, DR Mohan Raj, KE Dashiell and LEN Jackai). Co-publication of International Institute of Tropical and Japan International Research Centre for Agricultural Sciences, IITA, Ibadan, Nigeria

Pandita M L, Vashistha R N, Bhutani R D and Batra B R. 1982. Genetic variability studies in cowpea (*Vigna sinensis* (L.) Savi.)

under dry farming conditions. *Haryana Agricultural Univ. Journal of Research* 12: 241-245

Pant K C, Chandel K P S and Joshi B S. 1982. Analysis of diversity in Indian cowpea genetic resources. *SABRAO Journal* 14: 102

Pasquet R S. 1993a. Classification infraspecifique des formes spontanees de *Vigna unguiculata* (L.) Walp. a partir de donnees morphologiques. *Bull Jard Bot Nat Belg* 62: 127-173

Pasquet R S. 1993b. Two new subspecies of *Vigna unguiculata* (L.) Walp. (Leguminosae-Papilionoideae). *Kew Bulletin* 48: 805-806

Pasquet R S. 1997. A new subspecies of Vigna unguiculata (Leguminosae- Papilionoideae). Kew Bulletin 5: 840

Patel O P. 1973. Correlation studies in cowpea (*Vigna Sinensis* (L.). *JNKVV Research Journal* 7: 60-61

Pathak H C and Patel M S. 1992. Selection criteria in summer cowpea. *Gujrat Agricultural University Research Journal* 18: 113-114

Patil R B and Baviskar A.P. 1987. Variability studies in cowpea. *Journal of Maharashtra Agricultural University* 12: 63-66

Patil R B and Bhapkar D G. 1986. Combining ability in cowpea. *Journal of Maharashtra Agricultural University* 11: 303-306

Patil R B and Shete M M. 1987. Heterosis in crosses of seven genotypes of Cowpea. *Journal of Maharashtra Agricultural University* 12: 51-54

Patil S B and Patil R B. 1987. Genetics of some quantitative characters in cowpea. *Current Research Report MPAU* 36: 5-10

Pienaar B J and VanWyk A E. 1992. The Vigna unguiculata complex (Fabaceae) in southern Africa. *South African Journal of Botany* 58: 414-429

Pignone D, Cifarelli S and Perrino P. 1990. Chromosome identification in *Vigna unguiculata* (L.) Walp. Pages 144-150 In: *Cowpea genetic resources*, (Eds. N.Q. Ng and L.M. Monti), IITA, Ibadan, Nigeria

Ponmariamma T and Das L D V. 1997. Diallel analysis for yield and its components in cowpea *(Vigna unguiculata* (L.) Walp.). *Mysore Journal of Agricultural Sciences* 31: 7-11

Premsekar S and Raman V S. 1972. A genetic analysis of the progenies of the hybrid *V. sinensis* x *V. sesquipedalis*. *Madras Agriculture Journal* 59: 449-456

Premsekar S, Pornaiya BW and Raman V S. 1964. Studies on interspecific hybrids in Vigna (*V. sesquipedalis* x *V. sinensis*) and their derivatives. *Madras Agriculture Journal* 51: 361

Rachie K O. 1985. In: Cowpea Research, Production and Utilization (Singh, S.R. and Rachie, K.O. eds). Pp. 1-xxviii, Wiley Chichester

Radhakrishnan T and Jegaraj S. 1992. Genetic variability in cowpea. *Madras Agriculture Journal* 69: 216-219

Ram T, Ansari M M and Sharma T V R S. 1994. Relative performance of cowpea genotypes in rainfed conditions of Andaman and their genetic parameters analysis of seed yield. *Indian Journal of Pulses Research* 7: 72-75

Rawal K M. 1975. Natural hybridization among wild, weedy and cultivated *Vigna unguiculata* (L.) Walp. *Euphytica* 24: 699-707

Rawal K M, Rachie, K O and Frankowiak J D. 1975. Reduction in seed size in crosses between wild and cultivated cowpeas. *Journal of Heredity* 62: 253-254

Rewale A P, Birari S P and Jamadagni B M. 1995. Genetic variability and heritability in cowpea. *Agriculture Science Digest* 15: 73-76

Roquib, M A and Patnaik R K. 1990. Genetic variability in forage yield and components in cowpea (*Vingna unguiculata*). *Environment Ecology* 8: 236-238

Roy R S and Richharia 1948. Breeding and inheritance studies on cowpea (*Vigna sinensis*), *Journal of American Society of Agronomy* 40: 479-489

Saccardo F, Del Giudice A and Galasso I. 1992. Cytogenetics of cowpea. Pages 89-98. In: *Biotechnology enhancing research on tropical crops in Africa.* (Eds. G. Thottappilly, L.M. Monti, D.R. Mohan Raj and A.W. Moore). CTA/IITA Co-publication, IITA, Ibadan, Nigeria

Sahoo M S, Arora N D, Lodhi G P and Chandra S. 1971. Genotypic and phenotypic variability in cowpea (*V. sinensis*) under different environment conditions. *Journal of Research Punjab Agricultural University* 8: 159-164

Sakupwanya, S, Mithen, R, Mutangandura-Mhlanga, T. 1989. Studies on the African Vigna gene pool. II. Hybridization studies with *Vigna unguiculata* var. *tenuis* and var. *stenophylla. Plant Genetic Research Newsletter* 78-79: 5-9

Sangwan R S. 1996. Genetical and biochemical basis of yellow mosaic resistance in cowpea (*Vigna unguiculata* (L.) Walp.). Ph.D. Thesis, CCS Haryana Agricultural University, Hisar, India.

Sangwan R S and Lodhi G P. 1995. Heterosis for grain characters in cowpea (*Vigna unguiculata* (L.) Walp.). *Legume Research* 18: 75-80

Sangwan R S and Lodhi G P. 1998a. Genetics of seed weight in cowpea (*Vigna unguiculata* (L.) Walp.). *Forage Research* 24: 107-109

Sangwan R S and Lodhi G P. 1998b. Inheritance of flower and pod colour in cowpea (*Vigna unguiculata* (L.) Walp). *Euphytica* 102: 191-193

Sangwan R S and Lodhi G P. 1999. Inheritance of seed weight in cowpea (*Vigna unguiculata* (L.) Walp). *Indian Journal of Genetics and Plant Breeding* 59: 539-541

Sangwan R S and Lodhi G P and Bangarwa K S. 2000. Inheritance of seed coat colour in cowpea (*Vigna unguiculata* (L.) Walp). National Journal of Plant Improvement 2: In press

Sangwan R S and Lodhi G P and Kishor C. 1998. Inheritance of fodder yield and its traits in cowpea. *Haryana Agricultural University Journal of Research* 28: 91-94

Sangwan R S and Lodhi G P, Luthra Y P and Ahlawat T R. 1997. Inheritance of fodder protein content and in vitro dry matter digestibility in cowpea (*Vigna unguiculata* (L.) Walp). *Annals of Agricultural Biology Research* 2: 41-42

Sawant D S. 1994a. Association and path analysis in cowpea. *Annals of Agricultural Research* 15: 134-139

Sawant D S. 1994b. Gene control for yield and its attributes in cowpea. *Annals of Agricultural Research* 15: 140-143

Sawant D S. Biravi S P and Jadhav B B. 1994. Heterosis in cowpea. *Journal of Maharashtra Agricultural University* 19: 89-91

Selvi B, Rangaswamy P, Natarajan P, Ramalingam J and Vanniarajan C. 1994. Genetic variability in cowpea. *Annals of Agricultural Research* 15: 248-249

Sen N K and Bhowal J G. 1960. Cytotaxonomic studies on Vigna. *Cytologia* 25: 195-207

Sen N K and Bhowal J G. 1961. Genetics of *Vigna sinensis* (L.) Savi. *Genetica* 32: 247-266

Sharma C D and Singhania 1992. Perfomance of cowpea (*Vigna unguiculata* (L.) Walp) genotypes for fodder traits. *Annals of Arid Zone* 31: 65-66

Sharma P C, Mishra S N, Singh A and Verma J S. 1988. Genetic variation and correlation in cowpea. *Annals of Agricultural Research* 9: 101-105

Sharma P P and Joshi A K. 1993. Improvement of cowpea. Pages 255-275. In: *Advances in Horticulture* Vol. 5: Vegetable Crops: Part I (Eds. K L Chadha and G Kalloo), Malhotra Publishing House, New Delhi

Sharma T R. 1999. Genetic variability studies in cowpea. *Legume Research* 22: 65-66

Siddique A K M A R and Gupta S N. 1991b. Genotypic and phenotypic variability for seed yield and other traits in cowpea (*Vigna unguiculata* (L.) Walp). *International Journal of Tropical Research* 9: 144-148

Simmonds N W. 1962. Variability in crop plants, its use and conservation in cowpea. *Indian Journal of Genetics and Plant Breeding* 29: 104-109

Singh B, Khanna A N and Vaidya S M. 1964. Crossability studies in the genus Phaseolyus. JPG School 2: 47-50

Singh B B, Chambliss O L and Sharma B. 1997. Recent advances in cowpea breeding. Pages 30-49. In: *Advances in cowpea research.* (Eds. BB Singh, DR Mohan Raj, KE Dashiell and LEN Jackai). Co-publication of International Institute of Tropical Agriculture and Japan International Research Centre for Agricultural Sciences, IITA, Ibadan, Nigeria

Singh H B, Mittal S P, Dabas B S and Thomas T A. 1976. Breeder's stock of cowpea. *Indian Journal of Genetics and Plant Breeding* 36: 410-417

Singh K B and Jain R P. 1972., Heterosis and combining ability in cowpea. *Indian Journal of Genetics & Plant Breeding* 32: 62-66

Singh K B and Jindla L N. 1971. Inheritance of bud and pod colour, pod attachment and growth habit in cowpea. *Crop Science* 11: 928-929

Singh K B and Mehndiratta P D. 1969. Genetics variability and correlation studies in cowpea. *Indian Journal of Genetics and Plant Breeding* 29: 104-109

Singh Naveen, Singh V P and Singh J V. 1997. Genetic variability in cowpea (*Vigna unguiculata* (L.) Walp). *Forage Research* 23: 106-111

Singh R and Dabas B S. 1986. Inherence of yield and its components in cowpea (*Vigna unguiculata* (L.) Walp). *International Journal of Tropical Agriculture* 40: 85-87

Singh R, Singh S. and Paroda R.S. 1986. A study of heterosis heritability and genetic advances in cowpea. *Forage Research* 12: 43-47

Smartt J. 1985. Evolution of grain legumes III. Pulses in the genus Vigna. *Experimental Agriculture* 21: 87

Sree Kumar S G, Nair Y R, Saraswathy P, George M K and Thomas E J. 1979. Genetic variability and correlations in cowpea (*Vigna sinensis* (L.) Savi.) *Agricultural Research Journal of Kerala* 17: 227-231

Steele, W.M. 1976. Cowpeas, *Vigna unguiculata.* Pages 183-185. In: *Evolution of crop plants*, (Ed. N.W. Simmonds), Longmans, London, U.K.

Steele W M and Mehra K L. 1980. Structure, evolution and adaptation to farming system and environment in Vigna. Pages 393-404. In *Advance in legume science* (Eds. R.J. Summerfield and A.H. Bunting), HMSO, London

Thaware B L, Birari S P and Bendale V W. 1992. Genetic analysis in forage cowpea. *Annals of Agricultural Research* 13: 188-190

Thiyagarajan K, Natarajan C and Rathnaswamy R. 1989. Variability in Nigerian cowpea. *Madras Agricultural Journal* 76: 719-720

Tikka SBS, Jaimini S N, Asawa B M and Mathur J R. 1977. Genetic variability, interrelationship and discriminant function analysis in cowpea (*Vigna unguiculata* (L.) Walp). *Indian Journal of Heredity* 9: 1-9

Tikka SBS, Sharma R K and Mathur J R. 1976. Genetic analysis of flower initiation in cowpea (*Vigna unguiculata* (L.) Walp). *Zeitschrift fur Pflanzenzüchtung* 77: 23-29

Trehan K B, Bagrecha L R and Srivastava U K. 1970. Genetic variability and correlation in cowpea (*Vigna sinensis* (L.) Savi) under rainfed conditions. *Indian Journal of Heredity* 2: 39-43

Veeraswamy R, Palaniswamy G A and Rathnaswamy R. 1973. Genetic variability in some quantitative characters of cowpea (*Vigna sinensis* (L.) Savi.) *Madras Agricultural Journal* 60: 1359-1360

Venugopal R. 1998. Inheritance in cowpea (*Vigna unguiculata* (L.) Walp). Pod colour. *Crop Research (Hisar)* 15: 77-84

Venugopal R and Goud J V. 1977. Inheritance of pigmentation of cowpea. *Current Science* 46: 277

Verdcourt B. 1970. Studies in the Leguminosae-Papilionoideae for the flora of tropical East Africa IV. *Kew Bulletin* 24: 507-569

Viswanatha K P, Balaraju and Chakravarthy K K. 1998. Heterosis and inbreeding depression in cowpea, *Vigna unguiculata* (L.) Walp. *Mysore Journal of Agricultural Sciences* 32: 181-185

Zaveri P P, Patel P K and Yadavendra J P. 1980. Diallel analysis of flowering and maturity in cowpea. *Indian Journal of Agricultural Sciences* 50: 807-810

Zaveri P P, Patel P K and Yadavendra J P and Shah S M. 1983. Heterosis and combining ability in cowpea. *Indian Journal of Agricultural Sciences* 59: 793-796.

CHAPTER - 3

COWPEA GENETIC RESOURCES – A QUANTITATIVE PERSPECTIVE

S.K. Mishra

Cowpea (*Vigna unguiculata* L. Walp) is an important grain legume crop for the livelihood of the millions of rural poor in the underdeveloped/developing countries of the tropic and subtropic regions of the world. It is grown for a diversified uses for food, feed and green manuring. The young leaves, green pods and seeds are used as vegetable, while several snacks and meals are also prepared from the seeds. Cowpea assumes as an special advantage in the dryland farming as the luxurious vegetative growth and profuse root development help in soil moisture conservation and controlling soil erosion. It is mainly cultivated in Asia, Africa, Central and Southern America, and parts of Southern Europe and United States of America (Singh *et al.*, 1997). The countries like Bangladesh, China, India, Indonesia, Korea, Malaysia, Myanmar, Nepal, Pakistan, Philippines, Sri Lanka and Thialand are major cowpea growing countries in Asia (Mishra *et al.*, 1985). In India, it is grown all over the country. However, central and peninsular regions are the main cowpea growing areas in the country. The states like Rajasthan and adjoining parts of Gujarat, western Uttar Pradesh, Haryana, Punjab and parts of Himachal Pradesh represent the prominent areas of cowpea cultivation in north India.

TAXONOMY AND DISTRIBUTION

The cowpea belongs to genus *Vigna*, an amplified genus, comprising many pulse crops of the old world. Verdcourt (1970) considered this species containing two wild (*dekindiana, mensensis)* and three cultivated (*unguiculata, cylindrica, sesquipedalis)* forms. The taxonomic relationship among various species has been given in Table 1. Marechal *et al*., (1978) assigned all the cultivated forms of cowpea to sub-species *unguiculata*. However, Westphal (1974) classified it in four distinct cultivar groups viz., *unguiculata, biflora (=cylindrica), sesquipedalis* and *textilis*.

Table 1. Taxonomy of cowpea

Vigna unguiculata (L.) Walpers
ssp. *unguiculata* cv. gr. *unguiculata* cr. gr. *biflora* cr. gr. *sesquipedalis* cv. gr. *textiles*
ssp. *dekindtiana* (Harms) Verdcourt var. *dekindtiana* (Harms) Verdcourt
var. *mensensis* (Schweinf.) Marechal, Mascherpa & Stainer
var. *pubescens* (Wilczec) Marechal, Mascherpa & Stainer
var. *protracta* (E. Mey.) Verdcourt
ssp. *tenuis* (E. Mey.) Marechal, Mascherpa & Stainer
ssp. *stenophylla* (Harvey) Marechal, Mascherpa & Stainer

The cowpea has been recognized as of African origin. Marechal *et al.,* (1978) viewed that the distribution of wild forms covers much of tropical Africa, whereas the greater part of variability within the wild species is confined to south Africa. It has been emphasized that the cultivated forms, *cylindrica* and *sesquipedalis* have evolved in Asia while the subspecies *unguiculata* is prevalent in all the cowpea growing regions of the world.

ORIGIN AND DOMESTICATION OF COWPEAS

Faris (1965) suggested that the cultivated cowpea has evolved from the domestication of wild forms of *Vigna unguiculata dekindtiana* in west Africa. However, Stele (1976) opined the domestication from Ethiopia based on greater variability present in subspecies *dekindtiana* in Ethiopia and further dissemination could have occurred westwards across the landmass of Africa and eastwards across the Indian ocean (Hymowitz, 1972). Stele (1976) also suggested the diffuse origin of cowpea. Lush and Evans (1981) considered var. *Dekindtiana* as more probable progenitor of cultivated cowpeas than var. *mensensis*. However, both these forms (*dekindtiana, mensensis)* have different geopgraphical ranges. Ng and Marechal (1985) viewed that var. *dekindtiana* could have evolved from the crosses of var. *mensensis* (a very high degree of outcrossing tendency) and sub-species *unguiculata*. Since var. *mensensis* has more distinct morphological features as compared to cultivated types whereas var. *dekindtiana* has more or less intermediate features, the var. *mensensis* can be regarded as the most probable ultimate progenitor type. The richer diversity in Asia, generated mainly through much more disruptive selection of diverse uses, persuaded to recognized the Indian origin of cowpeas (Darlington, 1963).

Various kinds of forms have been evolved in cowpea over a period of time in different regions to cater the needs of specific uses. However, there is a general consensus on African origin of cultivar group *unguiculata*. These typical cowpeas are characterized by trailing growth habit and pods are generally held at the top of plant canopy. However, cowpeas were subjected to diverse range of selection pressures in Asia, particularly India, a major cowpea growing country. The *sesquipedalis* type cowpea might have been evolved through selection for long fleshy pods from the introduced *unguiculata* lines in the far East leading to their use as vegetable crop. In India, the cowpea were predominantly grown for animal fodder. Therefore, the introduced *unguiculata* lines were subjected to selection for a different kind of plant type. Such selections might have helped in the evolvement of erect and profuse growth habit with reduced number of nodes and internode length. However, the cowpeas are also used as vegetable and pulses in India, as evidenced by the development of fairly large number of varieties in the recent past for diversified uses. In fact all kind, viz., fodder, vegetable (green pod), grain type and dual purpose varieties are available for commercial cultivation. Thus, much diversity in cultivated cowpeas, evolved through differential selection pressures in different regions, may be the principal factor in recognizing various cultivar groups in cowpea (Table 2).

Table 2. Major characteristics of various cultivar groups of cowpea

Character	cv. gr. unguiculata	cv. gr. *biflora*	cv. gr. *sesquipedalis*	cv. gr. *textilis*	var. *mensensis*	var. *dekindtiana*	var. *pubescens*
Flower colour	white, purple	white, purple	white, purple	white, purple	purple	purple	purple
Standard petal							
width (mm)	24-30	24-28	28-35	24-27	33-42	23-42	28-32
length (mm)	18-23	18-24	21-24	19-21	-	17-34	19-21
Pod							
Length (cm)	6.5-25	7-13	15-90	7-14	-	6.0-11.6	-
Width (mm)	3-12	4-6	5-11	-	-	2.8-7.0	-
Orientation	Mostly pendant, vertical	Mostly pendant, vertical	All pendent	Vertical	Vertical	Vertical	Vertical
texture	fibrous, hard, firm; not inflated when young	fibrous, hard, firm; not inflated when young	succulent; inflated towards maturity; shrinking after maturity	fibrous, hard, firm, not inflated when young	fibrous, hard, firm, not inflated when young	fibrous, hard, firm, not inflated when young	fibrous, hard, firm, not inflated when young
dehiscence	nil	nil to moderate	nil	nil to moderate	shatters	shatters	shatters
Locules/pod	7-23	12-16	15.8-23.0	-	16-19	14-17	-
Calyx lobe (mm)	<5	<5	<5	<5	<5	<5	<5
Seed							
length (mm)	6-11	5-7	7-11	5.1-7.6	-	3-6	-
width (mm)	4-9	3-5	5-8	4.0-5.6	-	2-4	-
orientation in pod	crowded	crowded	crowded	crowded	crowded	crowded	crowded
Breeding system	inbreeder	inbreeder	inbreeder	inbreeder	outbreeder	inbreeder	inbreeder
Growth habit	erect, prostrate, climbing	prostrate, climbing	semi-erect, climbing	prostrate	prostrate, climbing	semi-erect, prostrate, climbing	semi-erect, prostrate
Shoot	glabrous	glabrous	glabrous	glabrous	glabrous	glabrous	pubescent

Source: Smart (1990).

GERMPLASM COLLECTION AND INTRODUCTION

The exploration and germplasm collection is an important activity of plant genetic resources (PGR) management. In early phase of pulse improvement programme, collections were made from the major pulse growing areas of the country and were evaluated for different desirable attributes like resistance to diseases and insect-pests, yielding ability and adaptability etc. The systematic efforts by the National Bureau of Plant Genetic Resources (NBPGR), New Delhi and its Regional Stations enabled the collection of indigenous cowpea germplasm from almost entire cowpea growing areas of the country resulting in the collection of 3843 accessions (Gautam *et al.*, 2000). Although major areas of the country have been covered, there are still several potential regions like Jammu & Kashmir, Assam, Meghalaya, Sikkim, Nagaland etc. which need to be explored vigorously.

As mentioned earlier, due to differential selection pressures in different regions, cowpea germplasm has a wide range of variation for almost all plant and seed characters like growth habit, plant height, stem pigmentation, leaf characters (color, shape, size, pigmentation, density), pod characters (length, breadth, shape color), seed characters (shape, size, colour, testa/seed coat pattern, eye-colour), yield potential and response to insect-pests and diseases.

The germplasm introduction from other countries mainly from Nigeria, Rhodesia and USA has played an important role in cowpea germplasm enrichment. A large number of introductions led to the identification of many useful donors for important economic traits. Many introductions with short plant height and long pods were made from Rhodesia. Similarly, photoperiod insensitive, early maturing and aphid resistant materials (EC216645, EC 216646, EC 216647, EC 216648) were introduced from USA. A large number of cowpea accessions were received from IITA, Nigeria. Some of them include resistance to aphids, bruchids and striga (EC 169716, EC 169717, EC 169718, EC 169719, EC 169720, EC 169721, EC 169722, EC 169723, EC 169724, EC 336589, EC 336590) and high yielding types (EC 149291, EC 149473, EC 149438). Several cowpea introductions have been identified with desirable attributes after proper evaluation (Table 3).

Table 3. Promising accessions of cowpea

Trait	Accession number
Early flowering	EC 390202, EC 390207, EC 390211, EC 390270, EC 390271, EC 90272, EC 390281, EC 390286, IC 202821, IC 202930, IC 209203, IC 209140
Early maturity	EC 240679, EC 240747, EC 240833, EC 240933, EC 244972, EC 244334, EC 267704, EC 367690, EC 390202, EC 390207, NIC 12496, NIC 15297 NIC 23006
Long pod	EC 367714, EC 390204, EC 390225, EC 390226, EC 390237, EC 390239, IC 199696, IC 199701, IC 202776, IC 202797, IC 202815, IC 202864, IC 202870, IC 209164, IC 209205
Erect plant type	EC 332349, EC 332351, EC 332358, EC 332359, EC 336899
Bold seeds	EC 244035, EC 244175
High yielding	NIC 22789, NIC 22794, NIC 22803, EC 309494, EC 343047, EC 367710

PLANT QUARANTINE

Evaluation of newly introduced germplasm, specially pulses, under quarantine tests is essential in order to detect and eliminate interception of insects and mites, fungus, virus and parasitic nematodes. A few interceptions made in cowpea germplasm are detailed below (Gautam *et al.*, 2000):

Interception	Name	Country
Insect & mites	*Anthonomus grandis*	USA, Zambia
	Bruchidius atrolineatus	Nigeria
	Callosebruchus chinensis	Nigeria, Philippines, USA
	C. maculates	Nigeria, USA
Plant pathogenic organisms	*Botryodiplodia theobromae*	Nigeria
	Cercospora graminicola	USA
Virus	Cowpea mottle virus	Philippines
	Unidentified (filamentous particles)	USA

GERMPLASM EVALUATION AND DOCUMENTATION

Proper evaluation and detailed characterization of germplasm is an important component of PGR management. To ensure uniform characterization of germplasm, the NBPGR has developed a set of well defined descriptors for pulses including cowpea. Earlier NBPGR had published three catalogues of cowpea based on evaluation of 707, 683 and 259 accessions for 34 and 24 descriptors, respectively (Singh and Rana, 1997). Recently, a catalogue based on evaluation of 424 accessions of cowpea for 34 discriptors has also been published (Sardana *et al.*, 2000). A wide range of genetic diversity has been observed for different traits in these germplasm (Table 4). In addition, about 1181 cowpea accessions including indigenous and exotic germplasm have been evaluated for 26 descriptors at the Division of Genetics, IARI, New Delhi (Sharma *et al.*, 1999). A wide range of diversity was observed for almost all the plant characters (Table 5). The range of expression recorded for different economic traits have been large. Based on multilocation testing, several promising accessions have been identified which can serve as potential donors for economic traits to supplement cowpea improvement programmes (Table 6).

Table 4. Variability for different traits in cowpea germplasm

Trait	Mean	Range
Days to first flowering	57.81	36-90
Days to 50% flowering	62.52	32-98
Days to first maturity	66.97	41-100
No. of primary branches/plant	7.24	3-13
Pod length (cm)	17.54	8.6-34
No. of pods/cluster	2.11	1-4
Days to final maturity	108.27	68-153
No. of seeds/pod	13.61	6-20
Seed yield/plant (g)	26.95	0.55-102
100-seed weight (g)	11.97	4.16-30

Table 5. Diversity for different characters in cowpea germplasm

Character	Type of diversity recorded
Growth habit	Erect, semi-erect, semi-spreading, semi-prostrate, prostrate
Growth pattern	Determinate, semi-determinate, indeterminate
Plant vigour	Very good, good, medium, poor
Pod pigmentation	Pigmented, mottled, tip pigmented, one side pigmented, non-pigmented
Immature pod colour	Green, dark green, light green, white, light yellow
Pod attachment to peduncle	Upright, straight, pendant
Pod curvature	Straight, medium curved, curved
Flower position	Terminal, non-terminal, throughout canopy
Leaf shape	Chordate, roundish, acute
Seed coat colour	White, buff, brown, black, dark brown
Seed coat pattern	Black, mottled, marbled, holstein, watson
Seed size	Large, medium, small, very small
Seed shape	Oval, globular, kidney shaped, rhomboid

Table 6. Promising cowpea accessions for economic traits

Trait	Accession number
Erect growth habit	C 16, C 33, C 73, C 79, C 85, C 287, C 858, C 859, C 863, C 876, C 883, C 892, C 916, C 920, C 980, C 985, C 994
Early flowering (< 40 days)	C 128, C 185, C 226, C 226, C 344, C 405, C 463, C 480, C 505, C 503, C 516, C 653, C 722, C 815, C 845, C 951, C 980, C 1049, C 1304
Early maturity (< 65 days)	C 102, C 405, C 480, C 556, C 653, C 722, C 777, C 792, C 815, C 845, C 873, C 951, C 980, C 1049, C 1051, C 1404
Long pod (> 20 cm)	C 222, C 572, C 601, C 615, C 617, C 626, C 642, C 649, C 853, C 887, C 941
Seeds/pod (> 15)	C 318, C 530, C 538, C 589, C 599, C 600, C 631, C 667, C 676, C 711, C 743, C 1037, C 1038, C 1270
Pods/peduncle (> 2.5)	C 23, C 330, C 356, C 400, C 405, C 421, C 472, C 440, C 606, C 653, C 687, C 695, C 705, C 849, C 854, C 865
100-seed weight (> 20 g)	C 12, C 447, C 485, C 525, C 526, C 602, C 605, C 607, C 636, C 851, C 924, C 1011, C 1024, C 1031, C 1186, C 1246, C 1258

GERMPLASM CONSERVATION

Due to rapid agro-ecological changes, fast erosion in genetic wealth has been witnessed in many crop plants. It is virtually not possible also to grow and maintain entire genepool in every crop season. Therefore, an effective approach is essential for conservation of the valuable germplasm including wild species, primitive cultures and land races. At present 2338 (1478 indigenous and 860 exotic) cowpea lines have been kept under long-term storage at -20^{0}C in the National Gene Bank (Dr. S. Sardana, Personal communication). Active or working germplasm accessions are maintained by the NBPGR headquarters and its Regional Stations, All India Coordinated Project and some of the volunteer centres like Division of Genetics, IARI, New Delhi. At present about 2200 accessions of cowpea are being maintained at NBPGR and its Regional Stations (Gautam *et al.*, 2000). About 800 cowpea germplasm accessions are also being maintained at the Division of Genetics, IARI, New Delhi.

UTILIZATION OF GERMPLASM

In early phase of varietal development, local cultivars with wider adaptability coupled with high yield and field resistance to major diseases were selected and released for commercial cultivation. In addition, the elite introduced germplasm after thorough evaluation were also released as variety or were used as parents in hybridization/inducing mutation in order to develop suitable varieties. The improved cultivars of cowpea developed through different breeding procedures have been given in Table 7. It is interesting to note that about 67% of the total cowpea varieties have been developed through selection from germplasm indicating a large scale utilization of indigenous and exotic materials in varietal development. However, the situation has not been encouraging in the hybridization as only eight (about 18%) varieties have been developed through this breeding procedure. The high rate of flower drop and poor success in crossing may be the reasons for non-utilization of hybridization technique in cowpea. Likewise, six varieties (V 16, V 37, V 38, V 130, V 240, V 585) were developed from IARI, New Delhi exploiting induced mutator system in a single cowpea variety Pusa Phalguni, an unique case of its kind in the world.

Table 7. Improved varieties of cowpea developed through different breeding methods

Breeding method		Total Number
Selection from exotic/ indigenous germplasm	Assem, Bundel Lobia 1, C 152, Chharodhi, CO3, CO1, CO4, CO5, Cowpea 78, Cowpea 88, FS 68, GC 1, GC 2, Gomti, JC 2, JC5, JC10, KBC 1, Lalitha, No. 2-1, No. 5-19-4-1, NP 3, Paiyur 1, PTB 1, Pusa Dophasali, Pusa Phalguni, Pusa Sawni, Rituraj, RS 9	29
Hybridization	Co2 (C 152 x C 149), Cowpea 74 (FS 68 x Cowpea No. 2), GC 82-7 (Black eye x V16), KM 1 (JC 5 x Pusa Sophasali), PTB 2 (P 118 x Kolinji Payar), RC 19 (Virginia x New era), S 288 (Sel. 2 x Virgina), S 488 (Virginia x Iran Grey)	8
Induced mutation	Amba (V16), Shreshta (V37), Swarna (V38), Rambha (V240), Pusa Safed (V130), Pusa Sampada (V585)	6
Total	—	43

FUTURE PERSPECTIVE

An intensive field evaluation of entire gemplasm available in cowpea needs to be undertaken for various economic traits including biotic and abiotic stresses. It is also essentially required to characterize the germplasm accessions at molecular (DNA) level in order to have precise information about real genetic worth/breeding value of a particular accession. Since India is considered to be one of the potential centres of origin of domesticated cowpeas, DNA profiling and fingerprinting will be useful in maintaining the correct identify of the potential materials, more so in the context of the recently introduced global IPR and WTO regimes. Some of the centres can be identified for molecular characterization of germplasm for specific trait (s) to accomplish the task as early as possible. The large-scale evaluation and utilization of wild species and geographic forms can help in broadening the genetic base of cowpea varieties. The germplasm enhancement and pre-breeding activities need to be promoted as an essential component to enhance the utilization of PGR in cowpea improvement. The necessary infrastructural facilities must be created for screening of germplasm under artificial epiphytotic conditions for major biotic and abiotic stresses to supplement the field screening. The inheritance and linkage relationships among traits of econmic importance must be given top priority to develop genetic linkage map. Publication of an up-to-date catalogue periodically will certainly help to have a database to identify desirable accessions. The registration of promising gremplasm should be encouraged to maintain a repository of valuable materials. Therefore, a multidisciplinary approach with well defined network for proper augmentation, evaluation, conservation and utilization of cowpea germplasm in needed to be pursued vigorously.

REFERENCES

Darlington, C. D. 1963. *Chromosome Botany and the Origins of Cultivated Plants.* Georage Allen and Unwin, London, U.K.

Faris, D. G. 1965. The origin and evolution of the cultivated forms of *Vigna sinesis. Can. J. Genetics & Cytology,* **7**: 433-452.

Gautam, P.L., Sharma, G.D., Srivastave, U., Singh, B.M., Kumar, A., Saxena, R.K. and Srinivasan, K. (eds.). 2000. *20 Glorious Years of NBPGR (1976-96).* National Bureau of Plant Genetic Resources, New Delhi. 333p.

Hymowitz, T. 1972. The trans-domestication concept as applied to guar. *Economic Botany,* **24**: 408-421.

Lush, W.M., and Evans, L.T. 1981. The domestication and improvement of cowpeas. *Evphytica,* **30**: 579-587.

Marechal, R., Mascherpa, J. M. and Stainier, F. 1978. Etude taxonomique d'un groupe complexe d'especes des genres Phaseolus et Vigna (Papilinaceae) sur la base de donnees morphologiques et polliniques, traittes par l' analyse f informatique. *Boissiera Geneve,* **28**: 1-273.

Mishra, S.N., Verma, J.S. and Jayasekara, S.J.B.A. 1985. Breeding cowpea to suit Asian cropping systems and consumer tastes. *In: Cowpea Research, Production and Utilization* (eds. S.R. Singh and K.O. Rachie). John Wiley & Sons, Chichester, U.K.: 117-123.

Ng, N.Q. and Marechal, R. 1985. Cowpea taxonomy, origin and germplasm. *In: Cowpea Research, Production and Utilization* (eds. S.R. Singh and K.O. Rachie). John Wiley & Sons, Chichester, U.K.

Sardana, S., Mahajan, R.K., Kumar, D., Singh, M. and Sharma, G.D. 2000. Catalogue on cowpea (*Vigna unguiculata* L. Walp.) germplasm. National Bureau of Plant Genetic Resources, New Delhi. 80p.

Sharma, B., Mishra, S.K., Tyagi, M.C. and Chaudhary, A.B. 1999. Cowpea genetic resources. Paper presented at *15th Group Meet of All India Coordinated Project on Arid Legumes*, 28-29 May, 1999, CAZRI, Jodhpur.

Singh, B. and Rana, R.S. 1997. Pulses germplasm collection and enhancement at NBPGR: Present accomplishment and future prospects. In: *Recent Advances in Pulses Research* (eds. A.N. Asthana and M. Ali). Indian Society of Pulses Research and Development, IIPR, Kanpur: 743-755.

Singh, B.B., Chambliss, O.L. and Sharma, B. 1997. Recent advances in cowpea breeding. *In: Advances in Cowpea Research* (eds. B. B. Singh, D.R. Mohan Raj, K.E. Danishiell, and L.E.N. Jackai). Copublication of IITA and JIRCAS. IITA, Ibadan, Nigeria: 30-49.

Smart, J. (ed.) 1990. Grain Legumes: Evoluation and Genetic Resources. Cambridge University Press, Cambridge, 379 p.

Stele, W.M. 1976. Cowpea. *In: Evolution of Crop Plants* (ed. N.W. Simmonds). Longman Press, London, U.K.: 183-185.

Verdcourt, B. 1970. Studies in the Leguminosae-Papilionoideae for the flora of tropical East Africa. IV. *Kew Bulletin,* **24**: 507-569.

Westphal, E. 1974. Pulses in Ethiopia: their taxonomy and agricultural significance. *In: Agricultural Research Report,* Centre for Agricultural Publishing and Documentation, Wageningen, the Netherlands.

CHAPTER – 4

COWPEA GENETIC RESOURCES - A Biotic Perspective

D.D. Saini and S.P.S. Choudhary

ABSTRACT

The distinguishing features of three sub species of cowpea have been described where, Asparagus bean is having longest pods and widely spaced seeds. All cultivated cowpeas are grouped under *V. unguiculata*, sub species *unguiculata* is sub divided into four cultigroups, namely Unguiculata, Biflora, Sesquipedalis and Textiles. Genetic factors for important desirable characters viz., pod length, pod filling, seed coat colour (governed by interaction of eight factors) and seed size have been reported. Further, the germplasm lines from 16 states of India evaluated for different characters where in, seeds/pod ranged from 2.0 to 43.0 and grain yield from 15.3 to 70.7 g/plant. At present, 3842 germplasm lines are available with NBPGR. In the source of resistance to fungal and bacterial diseases variety VITA-4 is resistant to Scab, Brown blotch, Septoria leaf spot and Bacterial blight and nine varieties (Tvu 39, Tvu 45, Tvu 99, Tvu 103, Tvu 1190, IT 82 82 D-889 & IT 83S-818) are resistant to cowpea yellow mosaic virus. Similarly, nine varieties (Tvu 108 P2, Tvu 410, Tvu 2740, Tvu 3417, Tvu 3509, EC-169718, EC-169723-24, ICV 11 & ICV 12) are resistant to aphids and two (Tvu 946 and Tvu 4557) to pod borer. Finally new sources/varieties developed in cowpea for different zones have been summarized.

INTRODUCTION

The ultimate success of any breeding programme aimed at developing superior genotypes or varieties for a particular agro-climatic region depends on the availability of diverse genetic base material or in other words the available genetic resources with the breeder. Thus probability of obtaining desirable plant types is directly proportional to the variability of available germplasm in almost all the crops and cowpea is no exception to this basic concept.

Hector (1936) described the characters of cowpea, catjang bean and asparagus bean. Verdcourt (1970) suggested that the seed type cowpea (formerly *V. sinensis (L.)* Savi ex Hassk), catjang or fodder type (formerly *V. cylindrica* (L.) Skeels), and the yard-long bean or asparagus bean, the vegetable type formerly (*V. sesquipedalis* (L.) Fruw. and *V. sinensis* (L.) Savi ex Hassk ssp. *sesquipedalis* (L.) Van Eseltine) are not to be regarded as distinct species since they cross freely to form fully fertile hybrids. He treated these three rather distinct taxa as subspecies of *V. unguiculata* (L.) Walp., i.e., cowpea = *V. unguiculata ssp. unguiculata,* catjang = *V. unguiculata* ssp. catjang and asparagus bean =*V. unguiculata* ssp. *Sesquipedalis.* The distinguishing features of the three subspecies are presented in Table-1. Verdcourt (1970) also recognized the wild or weedy type subspecies of cowpea (ssp. *dekindtiana* and ssp. *mensensis*), the progenitors of the cultivated types.

Later on the cultivated forms of cowpea Marechal *et al.* (1978) subdivided the genus described earlier by Verdcourt (1970) into seven subgenera. In this classification *V. unguiculata* (L.) Walpers and *V. nervosa* Markotter constitute the section *Catjang,* one of the six sections of the subgenus *Vigna.*

All cultivated cowpeas are grouped under *V. unguiculata,* subspecies *unguiculata,* which is subdivided into four cultigroups, namely Unguiculata, Biflora, Sesquipedalis and Textilis (Westphal 1974; Marechal *et al.* 1978; Ng and Marechal 1985).

The classification and nomenclature of the wild taxa within *V. unguiculata,* however, is complicated and could sometimes be confusing. An extensive work on characterization of 400 wild *V. unguiculata* accessions was conducted at IITA (Ng and Padulosi 1991; Padulosi 1993).

Table 1. The distinguishing features of the three subspecies of cowpeas (after Piper, 1912)

Character	Cowpea	Catjang bean	Asparagus bean
Pods flabby or inflated when dry	Neither flabby nor inflated	Neither flabby nor inflated	Inflated flabby shrinking between seeds before drying
Pod length (cm)	20-30	7.5-12.5	30-100
Pod position	Becoming pendent early .	Erect or asce-nding when green, spra-ding or refle-xed when dry	Pendent, flashy and brittle
Seed length (cm)	6-9	5-6	8-12
Seed shape	Subreniform to subglobose	Oblong or cylindric	Elongate kidney
Seed arrangement	Closely packed	Closely packed	Widely spaced

Within subspecies *protracta,* three varieties, namely var. *protracta* var. *rhomboidea* and var. *kgalagadiensis,* were distinguished. Similarly three varieties *tenuis, oblonga* and *parviflora* were recognized within the subspecies tenuis, while four new varieties namely var. *huillensis,* var. *congolensis,* var. *ciliolata* and var. *grandiflora,* have been proposed and added to the subspecies *dekindtiana.* Pasquet (1993a) proposed that the name subspecies *unguiculata* var. *spontanea* be used to describe all the weedy forms and the intermediates between truly wild var. *dekindtiana* and cultivated cowpea. The subspecies *burundiensis* (Pasquet 1993a) is a variant of var. *ciliolata.* It is found in mid-altitudes in Zaire, Burundi, Kenya and Uganda.

A. Genetic factors for important desirable characters in cowpea

1. **Pod length:** Capinpin and Irabagon (1950) observed that the F1 between long and short podded types was intermediate. There was skewness towards short pod in the F2 generation Yarnell (1965) reported that many

gene pairs were responsible for pod length. Roy and Richharia (1948) indicated that pod length was closely linked with fibrousness. Pod length is an important consideration in selection of vegetable types. It should be of medium size measuring more then 15 cm in length.

2. **Pod filling:** Close filling of pods by seeds will mean more number of seeds per pod x 100/pod length. This index gives a clearer picture of pod filling than the actual number of seeds per pod. Highest number of seeds per pod was noted in EC-4216 but the highest seed index was noted in EC-102159 (Singh *et al.*, 1976).
3. **Seed coat colour:** The seed coat colour in cowpea may spread uniformly over the whole seed or may be specked or marbled or may confine to a particular zone (Krishnaswamy, 1970)

Spillman and Sando (1930) described ten whole colour types (purple, black, dull black, blue, coffee, maroon, buff, red, pink and white) and proposed eight factors whose interactions produced these types. The factors were listed as: B-brown seed coat; F-very fine speckling giving rise to blue seed coat; N-anthocyanin pigment factor; P-purple seed coat; R-red seed coat (general factor for colour); T-less dense speckling characteristic of the Taylor variety; U-buff or clay coloured seed coat spotting; and X-taylor factor inhibitors.

According to Ebong (1970) the testa color is conditioned by three major genes and a host of modifiers. While seed color is controlled by a single recessive gene (Chambliss, 1973).

4. **Seed size:** Brittingham (1950) reported that seed size was quantitatively inherited. Seed size in the F1 of a cross between large seeded and small seeded parent exceeded that of the large seeded parent and in the F2 generation, seed size in many plants exceeded even that in the F1, while in the rest it approached the small seeded parent. Sene (1968) further suggested that six pairs of genes were involved in controlling seed size.

Wide range of genetic variability was observed in the germplasm collections from India for most of the morphological and agronomic characters. Most of the Indian cowpea material had prostate decumbent and erect cultivars, but the relative proportion of erect cultivars was low in the Indian material (Chandel and Pant, 1982; Chandel *et al.*, 1982)

683 germplasm lines from 16 states of India were evaluated for different characters (Table 2). Majority of the material from Gujarat, Rajasthan, Madhya Pardesh, Punjab, Himachal Pradesh and Uttar Pradesh were mainly *cylindrica* types or were intergrades between *cylindrica and unguiculata* while *unguiculata* types were more preponderant in Tamil Nadu, Kerala, Karnataka, Maharashtra and parts of Bihar. Exotic lines showed high degree of photoinsensitivity (NBPGR, 1987). Pod length varied from 10 to 38 cm, the former represents ssp. *cylindrica* and later ssp. *sesquipedalis*. The longest pod types are predominant in the material from Gujarat, Uttar Pradesh and Tamil Nadu. However, the smallest pod types were predominant in the collections from Himachal Pradesh. Pods per plant was the highly variable character and much influenced by day and night temperatures. However, production of pods in *kharif* season appeared to be more consistent than in the spring-summer (NBPGR, 1987).

Table 2. Variation for different characters in cowpea.

Character	Range	Reference
Days to flowering	39.0-97.0	Pant *et al.* (1982)
	48.0-95.0	Kumar & Mishra (1983)
Peduncle length (cm)	17.0-30.0	Kumar & Mishra (1983)
Pods per plant	10.0-35.0	Kumar & Mishra (1983)
Pods length (cm)	7.0-38.0	Pant *et al.* (1982)
	13.0-20.0	Kumar & Mishra (1983)
	10.0-38.0	NBPGR (1987)
Seeds per pod	2.0-43.0	Pant *et al.* (1982)
	14.0-18.0	Kumar & Mishra (1983)
100-seed weight (g)	4.0-24.0	Pant *et al.* (1982)
	8.0-17.0	Kumar & Mishra (1983)
	3.8-16.4	NBPGR (1987)
Grain yield per plant (g)	15.3-70.7	Kumar & Mishra (1983)

NBPGR was having a good collection (3539 lines) of indigenous accessions collected during the period 1976-1992 (Singh and Rana 1994). Presently the number of total lines available has further increased upto 3842 as reported at XVIth annual workshop held at S.K.Nagar, Gujarat during may, 2000.

B. Sources of resistance to diseases: Evaluation of germplasm have suggested that resistance is associated with collections from particular regions (Ahmad *et al.* 1972). Sources of resistance to fungal and bacterial diseases in cowpea are presented in Table 3. Cowpea germplasm has been evaluated for resistance to *Cercospora* leaf spot and cultivars resistant to *C. cruenta* have been identified (Verma and Patel 1969; Fery *et al.* 1976; Vakili, 1977; Williams, 1977a). Singh *et al.* (1976) reported that Barsati mutant and Sel 24/8-2 had field resistance to *Macrophomina* species.

Table 3. Sources of resistance to fungal and bacterial diseases in cowpeas

Disease/line	Reference
Fungal diseases	
Cercospora leaf spot	
Ala. 963.8, CR 17-1-34	Fery *et al.* (1975, 1976)
Fusarium wilt	
Iron, Victor, Brabham	Mackie (1934)
Charcoal rot	
Iron, Victor, Brabham	Mackie (1934)
Powdery mildew	
Chinito, No. 0199	Fennell (1948)
Culture 1-26	Satyanarayana (1994)
Phytophtora stem rot	
Ku 235, TVu 3861	Singh *et al.* (1982)
Collar rot	Viswanathan & Viswambharan (1979)
PLB-1, Calicut-78, selection 1476, C-57, Pusa Barsati	
Scab	
VITA-4	Allen *et al.* (1981a)
Brown blotch	
VITA-4	Allen *et al.* (1981a)
TVx-3236	Adebitan *et al.* (1992)
Septoria leaf spot	
VITA-4	Allen *et al.* (1981a)

Bacterial blight TVu 580, TVu 726, TVu 4540, TVu 410, VITA-3, VITA-4, TVx 1850-01E, IT90K-284-2, IT90K-277-2, IT86D-719, IT81D-1228-11 Mouride and Melakh	Allen *et al.* (1981a) Singh (1993)

Sources of resistance to viral diseases have been presented in Table 4. Resistance against blackeye cowpea mosaic virus have been identified by Lapido and Allen (1979b) and are being used in Nigeria. Immunity to cowpea severe mosaic virus (CSMV) has been identified in cv. Macailo (Lima and Nelson, 1977) and in the lines TVu 612, TVu 1460-2, TVu 1948 and TVu 2480. Five thousand cowpea lines were screened for field resistance to anthracnose, *Cercospora* leaf spot, rust and bacterial pustule (Williams, 1977a). After preliminary observations on 5000 lines, 719 lines were selected for further evaluation in field disease nurseries. All 719 lines were resistant to at least one of the four diseases, 685 lines were resistant to at least two of the four diseases, 537 lines were resistant to at least three, 208 were resistant to all four and of these 28 lines were also resistant to target spot. Sixteen of the multiple resistant lines were found resistant also to two isolates of cowpea yellow mosaic virus (IITA 1974a) Some cultivars are now available which combine resistance to as many as 10 diseases (Williams 1977a,b) for instance TVu 990.

The germplasm screening for disease resistance showed that India possesses a great potential. In all 153 lines were found completely free from cowpea yellow mosaic virus, bacterial leaf spot, *Cercospora* leaf spot and powdery mildew under field as well as controlled epiphytotic conditions (Patel and Jindal, 1970; Pant *et al.* 1982; NPPGR, 1987). Resistance genes were concentrated in *unguiculata* type (108 strains) as compared to *cylindrica* (48 strains) type.

Table 4. Sources of resistance to virus diseases in cowpea

Disease/line	Reference
Cowpea yellow mosaic virus TVu 39, TVu 45, TVu 99, TVu 103, TVu 106, TVu 1190 IT 82D-889, IT83S-818	Singh (1993), IITA (1974a)

Cowpea severe mosaic virus	
TVu 612, TVu 1460-2, TVu 1948, TVu 2480	Fulton and Allen (1982)
Cucumber mosaic virus	
Garbancito, Azul Grande, Chinito, No. 0199	Fennell (1948)
IT83D-442, IT85F-867-5	Singh (1993)
Blackeye cowpea mosaic virus	
Brown Crowder, Mangolia Blackeye, Mississippi Silver, Mississippi purple, Worthmore	Collins *et al.* (1985)
Cowpea banding mosaic virus	
Folden Type, Grey, EM2543, EC19739	Sharma and Verma (1981)
Cowpea aphid-borne mosaic virus	
TVu 134, TVu 222, TVu 1185, TVu 1258-1, TVu 1453, TVu 1630, TVu 2480,	Patel *et al.* (1982a)
IT82D-889, IT83S-818, IT83D-442, IT85F-867-5	Singh (1993)
Mouride	Cisse *et al.* (1995)

C. Sources of resistance to Insect-pests:

The identification of resistance to insect-pests in cowpea, began in the U.S.A. in the beginning of this century by Orton (1902) and Webber and Orton (1902) who noticed that cv.Iron was resistant to root knot nematode. Subsequent studies have led to the identification of other sources of resistance in U.S.A. (Hare 1959; Thomason and Mc Kinney 1960), Nigeria (Amosu, 1974; Ogbuji, 1978), India (Sharma and Sethi, 1976; Patel *et al.* 1977) and Brazil (Lemos and da Ponte, 1978).

Both antibiosis and non-preference (Karel and Malinga, 1980) mechanisms of resistance were responsible against the leaf-hoppers and pea aphids attack. Severe aphid mortality was observed as the cause of resistance (Singh, 1977).

Sources of resistance against some of the insect-pests are compiled in Table 5. Of these TVu 4557 (VITA 5) was resistant to leaf hoppers and pod borers. TVu 410 was resistant to aphids and pea aphids and TVu 1509 was resistant to leaf hoppers and thrips.

Table 5. Sources of resistance to different insect-pests in cowpea

Insect-pests/line(s)	Reference
Leaf hoppers	
TVu 59, TVu123, TVu 190, TVu 662, TVu 1509	Karel & Malinga (1980)
TVu 1190E TVu 662, TVu 3629, TVu 557 (VITA-5)	Raman *et al.* (1978)
Aphids	
TVu 108P2, TVu 410, TVu 2740, TVu 3417, TVu 3509	Singh (1977)
EC-169718, EC-169723-24	NBPGR (1987)
ICV-11, ICV-12 (mutants)	
Pea aphids	
TVu 408. TVu 410	Karel & Malinga (1980)
Thrips	
TVu 1509, TVx 2869P2-2	Singh (1977)
TVx 3343-01	Ehlers (1994)
Pod borer	
TVu 946, TVu 4557 (VITA 5)	Singh (1977)
TVu 946	Jackai (1982)
Pod sucking bug	
TVu 1977 (VITA 4), TVu 7274	Singh (1977)
TVx 3342-01	Ehlers (1994)
Cowpea curculio	
Fla. 453-01	Todd & Canerdy (1968)
Cowpea storage weevil	
TVu 2027,	Singh (1977)
RS 9	Dabi *et al.* (1979)
JC-5, 25/8/2/2, RS-9, CG 28, Sel 1476	Vir (1983)
EC 169716-17, EC 169720-22	NBPGR (1987)
Root knot nematode	
Iron, Victor, Brabham	Mackie (1934)
Bruchus sp.	
T2	Chandola *et al.* (1969)

D. Development of new sources/varieties in cowpea:

The concerted efforts were initiated about three decades back for the varietal development in cowpea and since then several varieties have been developed and released at central and state levels in this crop. The main objectives for breeding in cowpea today are development of high yielding, disease and insects-pest resistant, early maturing and white seeded varieties to cater the needs of diverse agro-climatic zones of the country. The cowpea is mainly cultivated as rainfed crop, so the early maturing varieties suit well to the farmers in the prevailing conditions and circumstances. According to the specific needs of different zones several varieties in this crop have been developed and recommended (Table 6).

Table 6. Recommended varieties of cowpea for different zones

S.No.	Zone	Varieties
1.	North-western zone (Punjab, Haryana; Western Uttar Pradesh, Rajasthan, Himachal Pardesh, Jammu and Kashmir)	C-152, Pusa Sawani, RC-19 (Rajasthan), cowpea-74. V-16, V-240, FS-68, T-2, V-130, RS-9, JC-5, JC-10, V-37, V-38, Pusa Phalguni, RC-101 (Raj.)
2.	North-eastern zone (Eastern Uttar Pradesh, Bihar, West Bengal, Orissa, Assam)	C-152, Pusa Sawani, Cowpea-74, V-16 (Amba), V-240, T-2, RC-19, Gomati, FGS-1
3.	Central zone (Madhya Pradesh, Gujarat, Maharashtra)	C-152, V-16, V-240, K-11, K-14, GC-1 (Gujarat cowpea-1), GC-2 (Gujarat cowpea-2), GC-3 (Gujarat cowpea-3), No. 21, V-585, Gu. No. 5-19-4-1
4.	Peninsular zone (Andhra Pradesh, Karnataka, Tamil Nadu, Kerala)	C-152, V-16, V-240, Co.1, Co.2, Co.3, Co.4, S-288, S-488, JC-5, KM-1, Krishna-mani, Kanakmani, KBC-1, KBC-2, TVX-944, C-182

REFERENCES

Adebitan, S. A., Ikiotun, T., Dashiell, K.E. and Singh, S.R. 1992. Use of three inoculation methods in screening cowpea genotypes for resistance to two *Colletotrichum* species. Plant Dis. 76: 1025-1028.

Ahmad, S.T., Magoon, M.L. and Mehra, K.L. 1972. Screening of a world collection of cowpea for field resistance to leaf spot and virus diseases, SABRAO Newsl., 4: 103-106.

Allen, D.J., Emachebe, A.M., Ndimande, B. and Ragi, J.A. 1981a. Identification of resistance in cowpea to diseases of the African savannas. *Trop. Agric,* 58: 267-274.

Amosu, J.O. 1974. The reaction of cowpea (*Vigna unguiculata* (L.) Walp.) to the root knot nematode (*Meloidogyne incognita)* in Western Nigeria. Niger. Agric. J., 11: 165-169.

Brittingham, W.H. 1950. The inheritance of date of pod maturity, pod length, seed shape and seed size in the southern pea (*Vigna sinesis*). Proc. Amer. Soc. Hort. Sci, 56: 381-388.

Capinpin, J.M and Irabagon, T.A. 1950. A genetic study of pod and seed characters in *Vigna.* Phillipines Agric., 3: 263-277.

Chambliss, O.L. 1973. A green seed coat mutant in southern peas: mode of inheritance and commercial potential. Hort. Sci., 8: 283.

Chandel, K.P.S, Joshi, B.S. and Pant, K.C. 1982. Genetic resources of *Vigna* species in India and there contribution in crop improvement. In: Advance in Cytogenetics and Crop Improvement (eds; R.B. Singh, R.M. Singh, B.D. Singh). Kalyani Publ, New Delhi. India, pp. 55-56.

Chandel, K.P.S, and Pant, K.C. 1982. Genetic resources of *Vigna* species in India, their distribution, diversity and utilization in crop improvement. Ann. Agric. Res., 3 (1-2): 19-34.

Chandola, R.P., Trehan, K.B. and Bagrecha, L.R. 1969. Varietal resistance of Bruchus sp. in cowpea (*Vigna sinensis)* under storage conditions, Curr. Sci., 38: 370-371.

Collins, M.H., Witchar, W., Barnett, O.W. and Ogle, W.L. 1985. Reactions of 16 cowpea cultivars to six viruses Pl. Dis., 69: 18-20.

Dabi, R.K., Gupta, H.C. and Sharma, S.K. 1979. Relative susceptibility of some cowpea varieties to pulse beetle, *Callosobruchus maculatus* (Fabricius). Indian J. Agric. Sci., 49: 48-50.

Ebong, U.U. 1970. A classification of cowpea varieties in Nigeria in subspecies and groups. Niger. Agric. J., 7: 5-9.

Ehlers, J.D. 1994. Correlation of performance of sole crop and intercrop cowpeas with and without protection from insect pests. Field Crops Research 36: 133-143.

Fennell, J.L. 1948. New cowpea resistance to mildew. J. Hered., 39: 275-279.

Fery, R.L., Dukes, P.D. and Jr. Cuthbert, F.P. 1975. Resistance in the southern pea (*Vigna unguiculata* (L.) Walp.) to *Cercospora* leaf spot. Hort. Sci., 10: 313.

Fery, R.L., Dukes, P.D. and Jr. Cuthbert, F.P. 1976. The inheritance of *Cercospora cruenta* leaf spot resistance in the southern pea (*Vigna unguiculata* (L.) Walp.) J. Amer. Soc. Hort. Sci., 101: 148-149.

Fulton, J.P. and Allen, D.J. 1982. Identification of resistance to cowpea servere mosaic virus. Trop. Agric., 59: 66-68.

Hare, W.W. 1959. Resistance to root-knot nematode in cowpea. Phytopathology, 49: 318 (Abstr.).

Harlan, J.R. 1971. Agricultural origins; centers and non centers. Science, 174: 468-474.

Hector, J.M. 1936. Introduction to the Botany of Field Crops II. Non-Cereals. Central News Agency Ltd. South Africa pp. 639.

IITA 1974a. International Institute of Tropical Agriculture Grain Legume Improvement Program. Report for 1973, Ibadan, Nigeria.

Jackai, L.E. 1982. Field screening technique for resistance of cowpea (*Vigna unguiculata)* to the pod-borer, *Maruca testulalis* (Geyer) (Lepidopotera: Pyralidae, Bull. Ent. Res., 72: 145-156.

Jeswani, L.M. and Baldev, B. 1990. Cowpea In: Advance in Pulse production technology. ICAR Publ., Krishi Anusandhan Bhawan, Pusa, New Delhi, India. Pp. 105-109.

Karel, A.K. and Malinga, Y. 1980. Leafhopper and aphid resistance in cowpea varieties. Trop. Grain Leg. Bull. 20: 10-11.

Krishnaswamy, N. 1970. Cowpea. In: Pulse Crops of India (ed,. P. Kachroo) Indian Council of Agril. Res. Krishi Bhawan, New Delhi, India, pp. 201-232.

Kumar, A. and Mishra, S.N. 1983. Genetic variability for seed yield and other traits in cowpea. Legume Res. 6: 83-85.

Ladipo, J.L. and Allen, D.J. 1979b. Identification of resistance to cowpea aphid-borne mosaic virus. Trop. Agric., 56: 253-259.

Lemos, J.W.V. and da Ponte, J.J. 1978. Cultivares de feijaadecorla, *Vigna sinesnsis* (L.) Savy, registentes a meloidoginose. Bol. Soc. Coaranese Agron., 19: 1-19.

Lima, J.A.A. and Nelson, M.R. 1977. Etiology and epidemiology of mosaic of cowpea in Ceara, Brazil. Pl. Dis. Repr. 61: 864-867.

Mackie, W.W. 1934. Breeding for resistance in blackeye cowpeas to *Fusarium* Wilt, charcoal rot and root knot nematode. Phytopathology 24: 1135 (Abstr.).

Marechal, R., Mascherpa, J.M. and Stainier, F. 1978. Taxonomy of cultivated *Vigna*. Boissiera 28: 1-273.

NBPGR, 1987. National Bureau of Plant and Genetic Resources. Achievements- A decade (eds; K.P.S. Chandel and B.M. Singh) New Delhi, India. Sci. Monogr. 11 pp 170.

Ng, N.Q. and Marechal, R. 1985. Cowpea taxonomy, origin and germplasm. In: Cowpea research, production and utilization, edited by S.R. Singh and K.O. Rachie. John Wiley and Sons, Chichester, U.K. pp. 11-21.

Ng, N.Q. and Padulosi, S. 1991. Cowpea genepool distribution and crop improvement. In: Crop genetic resources of Africa Vol. II edited by N.Q. Ng, P. Perrino, F. Attere and H. Zedan, IITA, CNR, IBPGR and UNEP, IITA, Ibadan, Nigeria. Pp 161-174.

Ogbuji, R.O. 1978. Reactions of cowpea to *Meloidogyne incognita* in Eastern Nigeria, PANS, 24: 32-34.

Orton, W.A. 1902. The wilt disease of the cowpea and its control U.S. Dep. Agri. Bur. Pl. Ind. Bull., 17: 9-20.

Padulosi, S. 1993. Genetic diversity, taxonomy and ecogeographic survey of the wild relatives of cowpea (*Vigna unguiculata* (L.) Walpers. Ph. D. thesis. Universite Catholique de Louvain-la-Neuve, Belgique.

Pant, K.C., Chandel, K.P.S. and Joshi, B.S. 1982. Analysis of diversity in Indian cowpea genetic resources. SABRAO J. 14: 103-111.

Pasquet, R.S. 1993a. Classification infraspecifique des formes spontaneeas de *Vigna unguiculata* (L.) Walp. (Fabaceae) a partr de donnees morphologiques. Bulletin du Jardin Botanique National de Belgique 62: 127-173.

Patel, G.J., Shah, H.M. and Patel, D.J. 1977. Screening of cowpea cultivars against root-knot nematodes, Indian J. Nemat; 7: 169-170.

Petel, P.N. and Jindal, J.K. 1970. Studies on resistance to bacterial diseases in India II. Resistance in cowpea to bacterial blight (canker). Indian Phytopath. Soc. Bull., 6: 28-34.

Patel, P.N. and Mligo, J.K., Leyna, H.K., Kuwite, C. and Mumbaga, E.T. 1982a. Sources of resistance, inheritance and breeding of cowpeas for resistance to a strain of cowpea aphid-borne mosaic virus from Tanzania. *Indian J. Genet. Plant Breed.*, 42: 230-239.

Piper, C.V. 1912. Agricultural Varieties of the Cowpea and Immediately Related Species. U.S. Dep. Agric. Bur. Pl. Ind. Bull., 229 pp. 160.

Raman, K.V., Singh, S.R. and van Emden, H.F. 1978. Yield losses in cowpea following leafhopper damage. *J. Econ. Entomol.*, 71: 936-938.

Roy, R.S. and Richharia, R.H. 1948. Breeding and inheritance studies in cowpea. *J. Amer. Soc. Agron.*, 40: 479-489.

Satyanarayana, A. 1994. Souvenir- 25 years of research on pulses in India. International Symposium on Pulses Research, April 1994, New Delhi, India pp. 133.

Sene, D. 1968. Inheritance of 100-seed weight in *Vigna unguiculata* Agron. Trop. (Paris). 23: 1345-1351.

Sharma, N.K. and Sethi, C.L. 1976. Reaction of certain cowpea varieties to *Meloidogyne incognita* and *Heterodera cajani.* Indian J. Nemat., 6: 99-102.

Sharma, S.R. and Verma, V. 1981. Reaction of some cowpea cultivars and lines against three sap transmissible viruses. Prog. Hort, 13(314): 127-129.

Singh, Dhan Pal 1991. Cowpea: In Genetics and breeding of Pulse crops. Kalyani Publishers, New Delhi: India: pp. 161-216.

Singh, S.R. 1977. Cowpea cultivars resistant to insect-pests in world germplasm collections: Trop. Grain Legume Bull., 9: 3-7.

Singh, B.B. 1993. Cowpea breeding. In: Archival report (1988-1992) of Grain Legume Improvement Program. IITA, Ibadan, Nigeria pp 10-53.

Singh, Bhag and Rana, R.S. 1994. Pulse germplasm collection and enhancement at NBPGR- Present accomplishments and future prospects. Presented at the Intl. Symp. on Pulse Crops Research and Development organized by Ind. Soc. Pulses R and D and ICAR, New Delhi., 2-4 April, 1994. pp.25.

Singh, B.B., Maligo, J.K., Marenge, E.T. and Ibrahim, I.A. 1982. Sources of resistance to *Phytophthora* stem rot in cowpea. Trop. Grain Leg. Bull., 25: 17-20.

Singh, H.B., Mittal, S.P., Dabas, B.S. and Thomas, T.A. 1976. Breeders stocks of cowpea. Indian J. Genet. Plant Breed., 36: 410-417.

Spillman, W.J. and Sando, W.J. 1930. Mendelin factors in the cowpea *(Vigna species).* Michigan Acad. Sci., Arts and Letters Papers, 11: 249-283.

Thomason, I.J. and McKinney, H.E. 1960. Reaction of cowpea. *Vigna sinensis* to root knot nematode, *Meloidogyne* spp. Pl. Dis. Reptr., 44: 51-53.

Todd, J.W. and Canerdy, T.D. 1968. Reaction of southern peas to the cowpea curculio. J. Econ. Entomol., 61: 1327-1329.

Vakili, N.G. 1977. Field screening of cowpeas for *Cercospora* leaf spot resistance. Trop. Agric., 54: 69-76.

Verdcourt, B. 1970. Studies in the leguminosae papilionoideae for the flora of East Tropical Africa. IV Kew Bulletin, 24: 507-569.

Verma P.R. and Patel, P.N. 1969. Host range, varietal resistance and epidemiological observations on *Cercospora* leaf spot disease of cowpea. Indian Phytopath., 22: 61-66.

Vir, S. 1983. Varietal resistance and susceptibility of cowpea to *Callosobruchus maculatus* Fab. Indian J. Ent., 45: 213-217.

Viswanathan, T.V. and Viswambharan, K. 1979. Screening germplasm collection of cowpea (*Vigna sinensis* Endl.) for tolerance/resistance to collar rot caused by *Rhizoctonia solani* Agric. Res. J. (Kerala). 17: 272-274.

Webber, H.J. and Orton, W.A. 1902. A cowpea resistant to root-knot *(Heterodera radicicola)* U.S. Dep. Agric. Bur. Pl. Ind. Bull., 17: 23-36.

Westphal, E. 1974. Pulses in Ethiopia: their taxonomy and agricultural significance. In: Agricultural research report. Center for Agricultural Publishing and Documentation, Wageningen, Netherlands. pp. 213-232.

Wiliams, R.J. 1977a. Identification of multiple disease resistance in cowpea. Trop. Agric., 54: 53-59.

Wiliams, R.J. 1977b. The identification of resistance to cowpea (yellow) mosaic virus. Trop. Agric., 54: 61-68.

Yarnell, S.H. 1965. Cytogenetics of the vegetable crops IV. Legumes. Bot. Rev., 31: 247-330.

CHAPTER - 5

COWPEA GENETIC IMPROVEMENT

S.K. Mishra, A.B. Choudhary and Daya Chand

ABSTRACT

Although cowpea is a minor crop of India, significant progress has been made in cowpea breeding. Several varieties with medium maturity, bold seeded types, semi-trailing growth habit with resistance to major diseases have been developed. Many potential donors for economic trails have also been identified which can be exploited for genetic improvement in yield potential of cowpea. Pyramiding genes for multiple resistance coupled with high yield need special attention. A systematic work must be initiated towards understanding the mecha-nism of drought tolerance identification of gene(s) conditio-ning drought tolerance and their molecular characterization. Efforts are also needed to develop saturated genetic linkage map of cowpea.

INTRODUCTION

Cowpea *(Vigna unguiculata* (L.) Walp.), popularly known as 'lobia', is an important legume crop grown for diverse uses such as grains for direct consumption used as processed foods, as vegetable (fresh leaves, green and tender pods), and as dry haulms and as fodder. Cowpea is often grown also as a cover crop for conserving soil moisture and soil improvement. The cowpea crop gives a luxurious vegetative growth and covers the ground to the extent that it effectively checks soil erosion in problem areas and can later be ploughed in as a source of green manure. Therefore, it holds considerable promise as pulse crop under dryland farming. Besides, it is a rich source of vegetable protein in combination with many other essential nutrients.

Trends in area, production and productivity

Cowpea is an important versatile food legume, grown in the tropics, subtropics of Asia, Africa, Central and Southern America and parts of the southern Europe and USA (Mishra *et al.*, 1985; Singh and Natre, 1985; Watt *et al.*, 1985; Heij, 1987; Fery, 1990; Hadjichristodoulou, 1992; Perrino *et al.*, 1993). It is difficult to estimate precise area and production of cowpea as it is generally grown as mixed crop and most of the countries do not maintain separate records on cowpea. Rachie (1985) reported that it occupied an area of about 9.0 million hectares in the world with a total production of 2.5 million tones of dry grain. Singh and Sharma (1996) reported that out of 9.0 million hectare area under cowpea, Africa alone accounts for 6.0 million hectares. Recently, Singh *et al.*, (1997) estimated that cowpea was grown in about 12.5 million hectares of area with an annual production of 3 million tones worldwide. The annual cropped area under cowpea in India is still difficult to estimate, as it is rarely grown as a pure crop. Cowpea is generally grown as a mixed crop with maize, sorghum, bajra, pigeonpea etc. Therefore, precise statistics on area and production of cowpea is not available in India. It has however been estimated that cowpea is cultivated in about 1.3 million hectares in Asia (Singh *et al.*, 1997), where India is a mojor contributing country.

The major cowpea producing countries include Nigeria, Niger Republic, Mali, Burkina Faso, Senegal, Ghana, Togo, Benin, Uganda, Tanzania, Zambia, Zimbabwe, Bostwana, Cameroon, Chad, Sudan, Somalia, Kenya, Malwi, Mozambique, India, Bangladesh, Nepal, Myanmar, Sri Lanka, Indonesia, China, Philippines, Brazil, Cuba, Haiti, USA, and the West Indies. In India, it is grown all over the country. However, western, central, and peninsular regions are the prominent areas of cowpea cultivation. In northern India the states like Rajasthan, Gujarat, U.P. Haryana, Punjab and parts of Himachal Pradesh are the main cowpea growing zones.

Current status of cowpea improvement activities in India

Although, cowpea is a minor pulse crop in India, considerable progress has been made towards the development of high yielding, disease resistant varieties (Table 1). As cowpea is grown for grain, fodder and vegetable purposes, the requirement of the variety for each category is different. The presentation of breeding efforts on fodder and vegetable cowpea is beyond the scope of the present article. The efforts related to improvement of grain cowpea have been described in the following sections.

Table 1. Promising cowpea varieties developed in India

Variety	Maturity duration (days)	Salient characteristic features	Average yield (kg ha^{-1})
C 152	105-110	Medium spreading, dark green foliage, pods pinkish at maturity, seeds buff and medium bold, field resistance to viral diseases, suitable for mixed cropping and a cover crop in plantation crops	1500-200
Type 2	125-130	Trailing type, dark green foliage, seeds bold and grey mottled	120-1800
Pusa Phalguni	110-115	Dwarf, bushy plant type, green foliage, white small seeds, susceptible to yellow mosaic virus (YMV)	1000-1200
Cowpea 74	120-125	Erect growth habit, dark green foliage, medium sized white seeds, resistant to viral diseases	1200-1500
Amba (V16)	90-100	Erect, non-trailing growth habit, dark green foliage, medium sized dark red seeds, field resistance to viruses, suitable for mixed cropping also	1000-1200

Rambha (V240)	90-100	Tall, trailing growth habit, dark green foliage, photo-period sensitive, medium sized dark red (brown) seeds, almost free to major cowpea diseases, highly adaptive with very high yield potential	1500-2000
V 38 (Swarna)	80-100	Erect, medium dwarf, moderately trailing, light green foliage, pods are held above the crop canopy due to long peduncles, seeds medium sized and bicoloured (fawn at hilum and red on dorsal surface), suitable for vegetable purpose also	1200-1500
GC 3	90-100	Tall, medium trailing, small thin pods, seeds with Watson pattern and medium sized, resistant to viral diseases	800-1000
Pusa Sampada (V585)	90-100	Tall, erect growth habit, light green foliage, medium bold creamy white seeds, field resistance to major cowpea diseases	800-1000

Major breeding objectives in cowpea

Most of the cowpea varieties developed so far are of trailing growth hahbit, luxurious vegetative growth and relative low yielding ability making the cowpea cultivation less remunerative. For taking a pure crop, the erect growth habit is the prerequisite so that the variety can respond to better management practices. Early or extra early maturity is another component of developing cowpea varieties so that they can fit well in popular cereal-based cropping systems for intensive agriculture.

Cowpea is severely attacked by a series of fungal (powdery mildew, rust, wilt, ashy stem blight, Anthracnose, web blight, scab, die-back etc.), bacterial blight and viral (cowpea aphid borne mosaic virus, cowpea yellow mosaic virus, cowpea golden mosaic virus, cowpea leaf curl virus and southern mosaic virus) diseases. Therefore, incorporation of disease resistance is essential for sustainable and stable cowpea production.

Recently, efforts have also been made on developing white seeded varieties of cowpea as it fetches higher prices in the market over the coloured seeded ones.

The cowpea is primarily grown under rainfed conditions where chances of crop failure are more due to moisture stress. Although early maturing varieties can escape the terminal drought, but they may also suffer badly under intermittent drought during vegetative phase (Singh, 1987). Thus, incorporation of drought tolerance will definitely boost the cowpea production.

Breeding methodology

Cowpea is a highly self-pollinated crop (Mackie and Smith, 1935; Williams and Chambliss, 1980), however, significant out-crossing occurs under certain environments primarily due to pollen transfer by the insects.

The cowpea is endowed with a very wide range of variability available for whole plant, pod and seed characters. This has helped in identification of promising genotypes simply by selection. A large number of varieties has been developed simply based on selection from local or exotic germplasm (Table 2).

The flowers of cowpea is mostly large and can be easily emasculated by hand and pollinated successfully with ease. Although effective hybridization procedures have been developed by many workers (Rachie *et al.*,1975; Blackhurst and Miller, 1980), very few number of varieties have been developed in India from hybridization (Table 2). A high degree of flower drop may be the possible reason for poor success of hybridization programme. Hybridization followed by pure line selection and progeny evaluation (pedigree method) has been the popular breeding method in cowpea improvement. The grain yields of advanced breeding lines (F_5 and F_6) were significantly correlated with grain yields of their respective F_3 lines, from which they were derived in cowpea trails conducted at IITA (Natare *et al.*, 1984) indicating the advantage of family selection in advanced generations.

Table 2. Cowpea varieties developed with different breeding methodologies in India

Breeding methodology	Variety
Selection	Bundel Lobia 1, C 20, C 152, Chharodhi, Co 1, Co 3, Co 4, Co 5, Cowpea 78, Cowpea 88, FS 68, Gomti, GC1, GC 2, GC 3, JC 2, JC 5, JC 10, K 11, KBC 1, Lalitha, No. 2-1, No. 5-19-4-1, No. 10, NP 3, Paiyur 1, Pusa Phalguni, PTB 1, RS 9, Type 2, Type 5269
Hybridization	Co 2 (C152 X C419), Cowpea 74 (FS68 X Cowpea No. 102), KM1 (JC X Pusa Dofasali), PTB 2 (P118 X Kolinji Payar), RC19 (Virginia X New Era), S 288 (Sel. 2 X Virginia), S 488 (Virginia X Iron Grey)
Induced mutation	V16 (Amba), V37·(Shrestha), V38 (Swarna), V130 (Pusa Safed), V240 (Rambha), V585 (Pusa Sampada)

Source: Modified from Sharma (1996).

Induction of a wide range mutator system was made possible in developing cowpea variety Pusa Phalguni which was found to alter almost all the plant and seed characters (Sharma, 1970). Many varieties (Amba, V 37, V 38, Rambha, V 130, Pusa Sampada) were developed through effective exploitation of this mutator induced variability at the Indian Agricultural Research Institute, New Delhi (Table 2). Pathak (1988) obtained two mutants (ICV11 and ICV 12) with 20 kR gamma rays irradiation to the parent variety ICV1. These mutants, in addition to differences in many other traits from the parent variety, were resistant to cowpea aphid. At IITA, the mutation breeding is currently being used to incorporate resistance against insects like pod borers, flower thrips, pod bugs etc.

Backcross method of breeding has also been advocated particularly for transferring the desirable genes from the exotic parents to the locally adopted genotypes (Ehlers and Foster, 1993). A cowpea variety CB 5 was developed in California using 4-6 backcrosses to the recurrent parent California Blackeye after an initial cross with the accession Iron which has resistance to root-knot nematode but many other undesirable feartures (Mackie, 1946). A few other varieties like CB 46, CB 88 (Helms *et al.*, 1991a, b) and Bettergro Balckeye (Fery. and Dukes, 1993) were also developed through employing backcross method of breeding. The transfer of dominant gene is relatively easy and quick in the backcross method as the backcross hybrid progeny can be detected easily and another backcross dose is given. However, when the traits are governed by recessive gene, the backcross progenies are selfed in each generation and screened for the presence of homozygous individuals carrying the recessive gene before giving additional backcross dose. The efficiency of backcross breeding can be enhanced using natural off-season nurseries or raising the materials in phytotron. The backcross breeding is likely to be more important breeding method as the gene(s) for economic traits from the exotic sources are discovered or unique sources are developed from attempting wide crosses.

The backcross method can also be effectively used in shuttle breeding programmes where two groups working in different conditions collaborate for developing varieties. Such kind of shuttle breeding is being used by IITA, Kano, Nigeria to incorporate resistance against Zakpota strain (Benin Republic) of striga into cowpea as the parasite cannot be brought into Nigeria.

The single seed descent (SSD) method, in which the seeds from each individual plants from F_2 and subsequent generations are collected to constitute bulk population, has not been widely used in cowpea breeding programmes. However, for effective breeding strategy, combination of pedigree, backcrossing and SSD method can be used depending upon specific breeding objectives.

The bulk method can be effectively used for population improvement programmes. The bulk method coupled with the mechanical sorting has been used for improving the seed size in cowpea (Redden *et al.*, 1984). The bulk method is also used for maintaining the genetic purity of breeder seed of the released varieties. This method can be also effectively employed for genetic purification of released varieties.

International status of cowpea breeding

A systematic cowpea breeding work is being pursued at the International Institute of Tropical Agriculture (IITA), Nigeria covering almost all the aspects of genetic improvement in cowpea. Several extra early varieties maturing in 60-70 days have been developed (Singh *et al.*, 1997) which can yield upto 25-28 q/ha. Several bushy vegetable varieties with semi-erect growth habit have also been developed (IT81D-1228-15 and IT86D-880) which can yield up to 18 t/ha green pods (Singh *et al.*, 1997). A number of high yielding varieties, suitable for intercropping, combining resistance to important insects pests and diseases have been developed. It was demonstrated that the early and medium maturing varieties performed better both under pure (sole) crop as well as inter crops. Several breeding lines have been developed with multiple resistance (Table 3) through systematic breeding work at IITA. These lines can serve as potential donors for transterring multiple resistance genes.

Table 3. Pyramiding genes for insect-pests and diseases resistance in cowpea

Pest/disease	Variety					
	Ife Brown	TVx 3236	IT82D-716	IT84S-2246	IT90K-59	IT90K-76
Aphid	S	S	S	R	R	R
Thrips	S	MR	MR	MR	MR	R
Bruchid	S	S	R	R	R	R
Striga	S	S	S	S	R	R
Alectra	S	S	S	S	R	R
Nematodes	S	S	S	R	R	R
Anthracnose	S	R	R	R	R	R
Cercospora	S	R	R	MR	R	R
Brown blotch	S	R	R	MR	R	R
Brown pustule	S	R	R	R	R	R
Bacterial blight	MR	MR	MR	MR	R	MR
Scab	S	MR	MR	MR	MR	R
Web blight	S	MR	MR	MR	MR	R
Yellow mosaic	S	S	R	R	R	R
Aphid borne mosaic	S	S	R	R	R	R
Golden mosaic	R	R	R	R	R	R

R= resistant; MR= moderately resistant; S=susceptible

Source: (Singh *et al.*, 1997)

Breeding for drought resistance has been an important area of cowpea research at IITA. A simple screening method has been devised using wodden boxes (Singh *et al.*, 1999) that accurately discriminates the resistant and susceptible cowpea varieties. Based on the days taken to permanent wilting and per cent recovery, cowpea lines IT 90K-59-2, Kanannado, Dan Ila, TVu 11979 and TVu 11986 were found to be drought tolerant and the lines TVu 7778 and TVu 8256 were drought susceptible (Singh *et al.,* 1999). Further studies demonstrated that two types of drought tolerance mechanisms were operating in cowpea (Mai-Kodomi *et al.*, 1999). In 'Type 1' mechanism, the drought tolerant lines (TVu 11986, TVu 11979) stop growth and conserve moisture in all the plant tissues and survive for over two weeks and gradually the entire plant parts dry out. On the other hand, in 'Type 2' mechanism, the drought tolerant lines (Dan Ila, Kanannado) continue slow growth of the trifoliates. However, with continued moisture stress, the trifoliates of these lines show early senescence and drop-off, however, the growing tips remain turgid and alive for even longer time. The nature of inheritance of both the types of drought tolerance mechanisms was worked out (Mai-Kodomi *et al.,* 1999b). The 'Type 1' and 'Type 2' mechanisms of drought resistance were found to be controlled by single dominant genes. However, the genes were different in the two types. These were designated as *Rds* (for 'Type 1') and *Rds 2* (for 'Type 2). The allelic test revealed that 'Type 1' (*Rds 1)* mechanisms was dominant over 'Type 2' (*Rds 2).* The monogenic dominant nature of drought tolerance in cowpea has opened a new vistas in understanding the genetic basis of drought tolerance. This will simplify our breeding efforts for drought tolerance as the same is comparatively easy to transfer and to screen the monogenic dominant traits in breeding programmes.

Besides, IITA is also working on quality aspects of cowpea e.g. protein content and shortening of cooking time. A wide range in protein content (22.9-32.5%) and cooking time (21.2. to 67.9 min) were observed among 100 cowpea lines evaluated at IITA. There are several reports on wide genetic variations for these traits in cowpea (Baker *et al.,* 1989; Nielsen *et al.,* 1993) indicating the possibility of combining these quality traits with high yield using appropriate breeding methodology.

Considerable progress has also been made in cowpea breeding in U.S.A (Fery, 1990). The breeding for heat and drought tolerance is the main focus of research at University of California, Riverside (Marfo and Hall, 1992). Breeding for disease resistance like viruses, *Fusarium* wilt and root-knot nematode are the major objectives. Most of the varieties developed in U.S.A. are horticultural types rather than grain types. Singh *et al.*, 1997 reviewed that about 23 varieties of cowpea have been released in U.S.A. during the last decade.

Opportunities for cowpea improvement in India

Cowpea is cultivated in India for various agricultural and domestic purposes. Therefore, varietal requirements for different purposes are variable making the cowpea breeding relatively complex. Some of the major areas for genetic improvement in cowpea are outlined below:

(a) Breeding for dwarf and erect growth habit

The cowpea is primarily grown as mixed or in intercropping combinations. The traditional varieties generally climb/trail on the main crop making the harvesting more cumbersome. It also has negative impact on grain yield. Therefore, development of cowpea varieties with dwarf and erect growth habit will help boost cowpea production under both mixed cropping or in sole cropping systems. Several dwarf and erect growing genotypes (C16, C 287, C 916, C 944) have been identified from large germplasm evaluation (Sharma *et al.*, 1999) which can be used as donors for developing suitable cowpea varieties for different agro-ecological regions (Table 4).

(b) Breeding for extra-earliness

Due to indeterminate growth habit, the cowpea plants keep on growing till later stages of crop growth leading to non-synchronous maturity. Therefore, efforts should be concentrated on developing early to extra-early maturing varieties (65-90 days) with synchronous maturity to suit under different cropping systems. Few genotypes which could mature within 70 days (C 780, C 653, C 722) have been identified at I.A.R.I., New Delhi and are being used in cowpea breeding programme (Table 4).

(c) Development of multipurpose varieties

As cowpea is grown for diverse purposes, it is, therefore, imperative to breed varieties which are suitable for grain, vegetable and fodder purposes. Although it is very difficult proposition, but taking the advantage of the faithfulness of genetic apparatus, it is possible to combine all the desirable features with marginal compromise.

(d) Breeding for resistance to diseases

Cowpea is attacked by about three dozen diseases caused by bacteria, fungi, viruses and nematodes. Among these yellow mosaic virus, golden mosaic virus, Anthracnose, Cercospora leaf spot and bacterial blight are the important diseases in Indian context. Several potential doners (Table 4) have been identified which can be used for pyramiding genes for disease resistance (Sharma *et al.*, 1999). The donors for multiple resistance developed at IITA (Table 3) can also be exploited in breeding programme for cowpea improvement.

Table 4. Potential donors for economic traits identified from cowpea germplasm

Traits	Donor lines identified
Erect growth habit	C 16, C 85, C858, C 892, C 916, C 973, C 994
Early maturity (<70 days)	C 480, C 653, C 722, C 792, C 873, C 980, C 1059, C 1307
Long pods (> 22cm)	C 222, C 573, C 615, C 626, C 642
Seeds per pod (> 15)	C 318, C 538, C 589, C 631, C676, C 1037, C 1270
Pods per peduncle (>2.5)	C 23, C 472, C 687, C 705, C 1096, C 1209, C 1308
100-seed weight (>20 g)	C 12, C 447, C 525, C 636, C 924, C 1031, C 1284, C 1306
YMV resistant	C 159, C 244, C 355, C 422, C 566, C 685, C 722, C 1018, C 1270, C 1300, C 1308
Leaf blight resistant	C 61, C 269, C 541, C 543, C 648
Leaf crinckle resistant	C 13, C 98, C 192, C 215, C 358
Rust resistant	C 38, C 76, C 370, C 425
Web blight resistant	C 10, C 11, C 49, C 139, C 140

(e) Breeding for insect resistance

About 85 insect species have been reported to attack cowpea (Booker, 1965). Of these, aphids, thrips and bruchids are common to damage the cowpea. Although pest management practices in cowpea have been advocated (Jackai and Adalla, 1997), the development of insect resistance varieties is the assured economic and eco-friendly way to control the insect incidence. The resistance sources of insects pests (Table 3) can be used for breeding insect resistant varieties.

(f) Breeding for tolerance to drought

Cowpea, not being a major pulse crop of India, not much breeding work has been attempted towards developing drought tolerant varieties. However, with the recent report regarding monogenic dominant control of drought tolerance in cowpea (Singh *et al.*, 1999; Mai-Kodomi *et al.*, 1999a, 1999b), it will be possible to develop drought tolerant varieties by simple hybridization and selection.

(g) Breeding for nutritional quality

Cowpea, one of the most important source of protein in the diets of tropical Africa, particularly in the Sahel region, is a higly nutritious crop with a dry seed protein content of about 25% with high digestibility (Table 5).

Table 5. Average protein content, protein digestibility and available proteins in cowpea

Cowpea species	Protein content (%)	Protein digestibility (%)	Available protein (%)
Wild	25.8	77.9	20.1
Cultivated	23.6	80.9	19.1

Source: Marconi *et al.*, 1990.

Besides, high protein content, cowpeas also have considerable amount of minerals, vitamins and calories. However, the cowpea seeds are low in sulphur containing amino acids (Bressani, 1985). Therefore, it is essential to develop varieties with higher protein content along with higher levels of sulphur amino acids. A judicious quality evaluation of advanced breeding materials can be integrated with the main breeding programmes to ensure the development of better quality materials.

Although, cowpea is free of toxic compounds, it also has many antinutritional factors (protease inhibitors, lectins, phenolic compounds, phytates, flatulence factors etc.) which not only affect the bioavailability of nutrients but also causes acceptability problems. However, the levels of some antinutritional factors have been considerably reduced in the cultivated genotypes, possibly, as a result of human selection (Table 6). It is, therefore, essential to pay due attention on quality parameters while developing the varieties. The sources (donors) available, particularly for high protein content and low cooking time can be exploited in systematic breeding programmes on cowpea improvement.

Table 6. Antinutritional factors in cowpea

Cowpea species	Trypsin inhibitor activity (TiV/mg)	Chymotrypsin inhibitor activity (CiV/mg)	Tanin content (mg/100 g.)
Wild	34.3	38.7	13.5
Cultivated	15.6	15.4	0.5

Sources: Marconi *et al.*, (1990).

Strategies for genetic improvement in cowpea

To meet the veritable objectives in cowpea improvement, following strategies can be adapted depending upon specific requirements and available facilites.

a. Augumentation, characterization and utilization of germplasm

Detailed characterization of germplasm is an integral part of crop improvement activities. Although few attempts have been made for characterization of germplasm of cowpea in India. However, the characterization has been mostly done for qualitative and quantitative traits based on multilocation testing. There is ample scope for germplasm evaluation mainly for identification of traits for resistance to biotic and abiotic stresses. There is need to develop reliable and efficient screening procedures of germplasm and breeding materials against various kinds of stresses. Application of DNA based biotechnological methods can be employed for germplasm characterization.

b. Utilization of wild and related species in cowpea improvement

The basic purpose of utilization of wild relatives and related species is to broaden the genetic base of the particular crop as the diversity is often reduced in the process of domestication. Several studies based on isozyme, seed proteins and chloroplast DNA have revealed that the genetic base of cultivated cowpea is narrow as compared to wild species.

Another important reason for exploitation of wild species is to transfer the traits of economic importance which are otherwise not available in cultivated cowpea e.g. resistance against pod borer, pod sucking bugs etc. Ng (1990) has reported that fertile hybrids can be obtained by crossing cultivated cowpea (*Vigna unguiculata* spp. *unguiculata)* with wild relatives belonging to *V. unguiculata* spp. *dekindtiana* (Harms) Verdc., spp. *stenophylla* (Harvey) M. M. & S., and spp. *tenuis* (E. Mey.) M. M. & S. A related cowpea species *Vigna vexillata* (L.) A. Richard, has been found to possess substantial resistance against pod sucking bugs, pod borer, and some other insects (Singh *et al.*, 1992). However, successful crosses with cultivated species with *Vigna vexillata* have not been reported. At IITA, the crosses between wild species (*Vigna unguiculata* spp. *dekindtiana)* and cultivated cowpea have been made in order to transfer resistance against pod borer and pod sucking bugs (Ng, 1992). The embryo rescue technique can be used, in some cases, for obtaining successful hybrids between cultivated and wild species (Fatokun and Singh, 1987).

c. Exploitation of heterosis in cowpea improvement

Occurrence of considerable heterosis for many traits including grain yield observed in cowpea (Ismail *et al.*, 1994; Menendez and Hall, 1996; Fery and Singh, 1997) has led the possibility of exploitation of hybrid breeding. Genetic male sterility reported in cowpea (Rachie and Rawal, 1976; Hall, 1992) can also favour the outcrossing. However, there are no reports on developing hybrids in cowpea.

d. Genetic studies in cowpea

Although cowpea offers a great opportunity for genetic studies by virtue of the wide range of variability available for various plant, pod and seed characters, very little efforts have been made to understand the mode of inheritance and linkage relationships. Development of genetic linkage map helps in deciding the precise location of gene(s) on specific chromosome. The linkage map involving all possible markers will definitely increase the efficiency of breeding as in some cases the marker based selection for specific disease resistance is difficult when the expression of resistance is influenced by the environment. Indirect selection for other morphological trait tightly linked to resistance can increase the selection efficiency. Therefore, development of comprehensive genetic linkage map based on morphological as well as molecular markers is essential for cowpea improvement.

e. Utilization of biotechnological tools in cowpea improvement

Recently developed biotechnological techniques offer a great promise in cowpea improvement. Identification of molecular markers linked to the traits of interest e.g. resistance to diseases and insect-pests, tolerance to abiotic stresses (like drought, heat and salinity) and quality parameters can help in marker-aided selection in an environment free manner. The indirect selection through the marker in early segregating population can help in the identification of the individual plants carrying the trait of interest at initial stages of the crop growth. Although linkage maps have been developed involving RFLP, RAPD, and AFLP markers but only two or three morphological markers have been mapped. For effective utilization of linkage map in breeding programmes it is essential to develop saturated linkage map with all possible traits of economic importance.

The cloning of important genes, like recently reported drought tolerance gene may revolutionize the breeding for tolerance to this stress. Subsequently, the cloned gene can also be transferred to other crop species suffering from drought. The process of genetic engineering can make significant strides in this context. Development of trasgenics particularly for insect resistance is another potential area for utilization of biotechnological tools. However, the regeneration protocol needs to be developed for the success of transgenic production. The microprojectile bombardment is another option to produce transgenics.

REFERENCES

Baker, T.A., Nielsen, S.S., Shade, R.E. and Singh, B.B. 1989. Physical and chemical attributes of cowpea lines resistant and susceptible to *Callosobruchus maculatus* (F.). *J. Stored Products Research,* 25: 1-8.

Blackhurst, H.T. and Miller, J.C. (Jr.) 1980. Cowpea. *In:* Hybridization of Crop Plants (eds. W.R. Fehr and H.H. Hadley). ASA and CSSA, Madison, WI, pp. 327-337.

Booker, R.H. 1965. Pests of cowpeas and their control in northern Nigeria. *Bulletin of Entomological Research,* 55: 663-672.

Bressani, R. 1985. Nutritive value of cowpea. *In:* Cowpea Research, Production and Utilization (eds. S.R. Singh and K.O. Rachie) John Wiley and Sons, U. K. pp. 353-359.

Ehlers, J.D. and Foster, K.W. 1993. Introgression of agronomic characters from exotic cowpea germplasm into blackeye bean. *Field Crops Res.,* 35: 43-50.

Fatokun, C.A. and Singh, B.B. 1987. Interspecific hybridization between *Vigna pubescens* and *Vigna unguiculata* (L.) Walp. through embryo rescue. *Cell Tissue Organ Cult.,* 9: 229-233.

Fery, R.L. 1990. The cowpea: production, utilization and research in the United States. *Horticultural Reviews,* 12: 197-222.

Fery, R.L. and Dukes, P.D. 1993. 'Bettergro Blackeye' southernpea. *HortScience,* 19: 456-457.

Fery, R.L. and Singh, B.B. 1997. Cowpea genetics: a review of recent literature. *In:* Advances in Cowpea Research (eds. B. B. Singh, D. R. Mohan Raj, K. E. Dashiell and L. E. N. Jackai), IITA/JIRCAS. IITA, Ibadan, Nigeria, pp. 12-29.

Hadjichristodoulou, A. 1992. Evaluation of cowpea (*V. unguiculata)* lines for dry seeds. Technical Bulletin No. 134. Cyprus Agricultural Institue, Nicosia, Cyprus.

Hall, A.E. 1992. Breeding for heat tolerance. *Plant Breed. Rev.,* 10: 129-168.

Heij, G. 1987. Expression in research with yardlong bean: a crop with prospects (in Netherlands). *Groente en Fruit,* 42: 44-45.

Helms, D.L.; Panella Buddenhagen, I.W.; Tucker, C.L.; Foster, K.W. and Gepts, P. L. 1991b. Registration of "California Blackeye 46" cowpea. *Crop Sci.,* 31: 1703.

Helms, D.L.; Panella Buddenhagen, I.W.; Tucker, C.L.; Foster, K.W. and Gepts, P.L. 1991a. Registration of "California Blackeye 88" cowpea. *Crop Sci., 31*: 1703-1704.

Ismail, A.M.; Hall, A.E.; and Bray, E.A. 1994. Drought and pod size effects on transpiration efficiency and carbon isotope discrimination of cowpea accessions and hybrids. *Austral. J. Plant Physiol.*, 21: 23-35.

Jackai, L.E.N. and Adalla, C.B. 1997. Pest management practices in cowpea: a review. *In:* Advances in Cowpea Research (eds. B.B. Singh, D. R. Mohan Raj, K.E. Dashiell and L.E.N. Jackai), IITA/JIRCAS. IITA, Ibadan, Nigeria, pp. 240-258.

Mackie, W.W. 1946. Blackeye beans in California. Univ. California Agr. Expt. Sta. Bulletin 696, California.

Mackie, W.W. and Smith, F.L. 1935. Evidence of field hybridization in beans. *J. Am. Soc. Agron.*, 27: 903-909.

Mai-Kodomi, Y.; Singh, B.B.; Myers, O.; Yopp (Jr.), J. H.; Gibson, P. J. and Terao, T. 1999a. Two mechanisms of drought tolerance in cowpea. *Indian J. Genet.*, 59(3): 317-323.

Mai-Kodomi, Y.; Singh, B. B.; Terao, T.; Myers, O.; Yopp (Jr.), J.H.; and Gibson, P.J. 1999b. Inheritance of drought tolerance in cowpea. *Indian J. Genet.*, 59(3): 317-323.

Marconi, E.; Lombardi-Baccia, G.; Carnovale, E. and Ng, N.Q. 1990. Nutritional evaluation of wild and cultivated species of cowpea. *In:* Cowpea Genetic Resources (eds. Ng, N.Q. and L.M. Monti), IITA, Nigeria, pp. 101-110.

Marfo, K. O. and Hall, A. E. 1992. Inheritance of heat tolerance during pod set in cowpea. *Crop Sci.*, 32: 912-918.

Menendez, C. M. and Hall, A. E. 1996. Heritability of carbon isotope discrimination correlations with harvest index in cowpea. *Crop Sci.*, 36: 233-238.

Mishra, S.N.; Verma, J.S. and Jayasekara, S.J.B.A. 1985. Breeding cowpea to suit Asia cropping systems and consumer traits. *In* Cowpea Research, Production and Utilization (eds. S. R. Singh and K.O. Rachie). John Wiley and Sons, U.K. pp. 193-197.

Natare, B.R.; Redden, R.J.; Singh, B.B. and Aken'Ova, M.E. 1984. Evaluation of early generation selection procedures for yield in cowpea (*Vigna unguiculata* L. Walp.). *Field Crops Res.*, 9: 91-100.

Ng, N.Q. 1990. Recent developments in cowpea germplasm collection conservation and research at Genetic Resources Unit, IITA. *In* Cowpea Genetic Resources (eds. Ng, N.Q. and L.M. Monti), IITA Nigeria, pp. 13-28.

Ng, N.Q. 1992. Wide crosses of *Vigna* food legumes. *In:* Biotechnology Enhancing Research on Tropical Crops of Africa (eds. G

Thottappilly, L. M. Monti, D. R. Mohan Raj and A. W. Moore). CTA/ IITA, IITA, Ibadan, Nigeria. Pp. 75-80.

Nielsen, S.S.; Brandt, W.E. and Singh, B.B. 1993. Genetic variability for nutritional composition and cooking time of improved cowpea lines. *Crop Sci.*, 33: 469-472.

Pathak, R. S. 1988. Induced mutation for resistance to aphid (*Aphis craccivora koch)* in cowpea. *In:* Proc. Workshop on Improvement of Grain Legume Production Using Induced Mutation, July 1-5, 1986, Pullman, WA, IAEA, Vienna. pp. 279-291.

Perrino, P.; Laghetti, G.; Zeuli, P.L.S. and Monti, L.M. 1993. Diversification of cowpea in the Mediterranean and other centres of cultivation. *Genetic Resources and Crop Evolution,* 40: 121-132.

Rachie, K.O. 1985. Introduction. *In:* Cowpea Research, Production and Utilization (eds. S.R. Singh and K.O. Rachie) John Wiley and Sons, U. K. pp. xii-xxviii.

Rachie, K.O. and Rawal, K.M. 1976. Integrated approaches to improving cowpeas *(Vigna unguiculata* L. Walp.). Technical Bulletin No. 5, IITA, Ibadan, Nigeria.

Rachie, K.O. and Rawal, K.M. and Franckowiak, J. D. 1975. A rapid method of hand crossing in cowpeas. Technical Bulletin No. 2, IITA, Ibadan, Nigeria.

Redden, R.; Drabo, I. And Aggarwal, V. 1984. IITA programme of breeding cowpeas with acceptable seed types and disease resistance for West Africa. *Field Crops Res.*, 8: 35-48.

Sharma, B. 1970. Induction of a wide range mutator in *Vigna sinensis.* Proc. BARC Symposium on Use of Radiation and Radiomimetic Substances in Plant Breeding, Bombay. pp 13-21.

Sharma, B. 1996. Other pulses. *In:* 50 Years of Crop Science Research in India (eds. R. S. Paroda and K. L. Chadha), ICAR, New Delhi. Pp. 297-317.

Sharma, B.; Mishra, S.K.; Tyagi, M.C.; and Chaudhary A.B. 1999. Cowpea genetic resources. Paper presented at 15th Group Meet of All India Coordinated Project on Arid Legumes, 28-29, May, 1999, CAZRI, Jodhpur.

Singh, B. B. Breeding cowpea varieties for drought escape. 1987. *In:* Food Grain Production in Semi-arid Africa (eds. J. M. Menyonga, T., Bezunen, and A. Youdeowei), OVA/STRC-SAFGRAD, Ouagadougou, Burkina Faso, pp. 299-306.

Singh, B. B.; Chambliss, O. L. and Sharma B. 1997. Recent advances in cowpea breeding. *In:* Advances in Cowpea Research (eds. B. B.

Singh, D.R. Mohan Raj, K.E. Dasniell, and L.E.N. Jackai). IITA/ JIRCAS, IITA, Ibadan, Nigeria, pp. 30-49.

Singh, B.B.; Mai-Kodomi, Y. and Terao, T. 1999. A simple screening method for drought tolerance in cowpea. *Indian J. Genet.*, 59(2): 211-220.

Singh, B.B. and Natare, B.R. 1985. Development of improved cowpea varieties in Africa. *In:* Cowpea Research, Production and Utilization (eds. S.R. Singh and K.O. Rachie). John Wiley and Sons, U. K. pp. 105-115.

Singh, B.B. and Sharma, B. 1996. Restructuring cowpea for higher yields. *Indian J. Genet.*, 56(4): 389-405.

Singh, S.R.; Jackai, L.E.N.; Thottappilly, G.; Cardwell, K.F. and Meyers, G.O. 1992. Status of research on constraints to cowpea production. *In:* Biotechnology : Enhancing Research on Tropical Crops of Africa (eds. G. Thottappilly, L.M. Monti, D.R. Mohan Raj and A. W. Moore). CTA/ IITA, IITA, Ibadan, Nigeria. pp. 21-26.

Watt, E.E.; Kueneman, E.A. and de Araujo, J.P.P. 1985. Achievements in breeding cowpea in Latin America. *In:* Cowpea Research, Production and Utilization (eds. S.R. Singh and K.O. Rachie) John Wiley and Sons, U.K. pp. 125-128.

Williams, C.B. III. and Chambliss, O.L. 1980. Outcrossing in southernpea. *Hort. Science,* 15: 179.

CHAPTER - 6

MUTATION BREEDING IN COWPEA

R.N. Pandey and S.E. Pawar

ABSTRACT

Mutation breeding research in cowpea (*Vigna unguiculata* (L.) Walp.) was initiated in India in the late sixties. Chemical mutagens were the first to be used for inducing mutations. DMS at 0.1-0.8%, EMS at 0.25-0.40% and NMU at 0.01-0.08% with exposure period of 2-16 h were used for induction of mutations. The physical mutagens, especially gamma rays at the dose range of 200-500 Gy were used in various studies. Differential radio-sensitivity among the cowpea genotypes has been observed. The useful dose for gamma rays appears to be below 250 Gy. Mutations affecting various morphological characters such as growth habit, seed size and colour have been induced by both physical and chemical mutagens. The activation of transposable elements in cowpea has been observed following radiation and chemical mutagenic treatments. Six mutant varieties have been released in cowpea for cultivation in India. Development of extra-early, dwarf and bushy plant type and drought tolerant varieties have been identified as major objectives for the improvement of cowpea. Several promising mutant lines already obtained through induced mutation are being evaluated in the state and national trials and also used in cross breeding programmes. Thus, mutation breeding can effectively supplement the conventional breeding approach in achieving the objectives for cowpea improvement.

INTRODUCTION

Cowpea (*Vigna unguiculata* (L.) Walp.) is an important legume of the tropics and subtropics of Asia, Africa, Central and South America, parts of southern Europe and United States of America (Singh and Sharma, 1996). With its various uses as grain, vegetable and fodder, cowpea is grown in several countries of the world. It has been estimated that cowpea is cultivated in almost 12.5 million ha with an annual production of about 3.0 million tonnes worldwide. It plays a major role in the livelihoods of millions of people living in less developed countries of the tropics. Central and west Africa account for over 64% of the area under cowpea cultivation in the world. A substantial part of the cowpea production comes from the drier regions of northern Nigeria. In most Asian countries, cowpea is a minor pulse crop. The cowpea growing countries in Asia are India, Sri Lanka, Bangladesh, Myanmar, China, Korea, Thailand, Indonesia, Nepal, Pakistan, Philippi-nes and Malaysia. The average yield of cowpea is low due to several biotic and abiotic factors as well as low yield potential of traditional varieties. Sporadic research efforts have been in progress since long in several countries (Singh and N'tare, 1985), cowpea improvement programme received greater attention worldwide from 1970 onwards, when Inter-national Institute of Tropical Agriculture initiated systematic breeding programme in collaboration with national progra-mmes of several countries. Breeding objectives for cowpea in recent years have been to combine high yield with upright growth habit, bushy determinate plant type and early maturity which can fit well into the cereal based cropping system. Evolving varieties with white bold seeds for consumers' acceptance, and green seed for processing purposes has also assumed considerable importance. Breeding varieties resistant to biotic and abiotic stresses has ever been the matter of prime concern. Efforts are therefore, going on in various countries to achieve some of these objectives.

Cowpea cultivation and breeding programmes in India

India alone accounts for half of the total acreage of about 1.3 million ha under different forms of cowpea in Asia (Singh *et al.* 1997). Nevertheless, cowpea is considered a minor pulse crop in India, and is cultivated in the semi-arid regions of Rajasthan, Gujarat, Karnataka, Tamil Nadu and Maharashtra mostly as grain legume. It is largely grown as *kharif* crop, but also cultivated as a second crop in *rabi* after paddy. Data on the total production of cowpea in the country are not available at present (Singhal, 1999). The crop has a great potential, as it can be grown in kharif and summer in north India and throughout the year in peninsular India (Jeswani and Baldev, 1990). The role of cowpea is predominantly that of a pulse, but after the introduction of 'Unguiculata' form (grain type) to India and south east Asia, two other culti-groups, 'Sesquipedalis', (vegetable type) and 'Biflora' (fodder type) were evolved under the human selection (Smartt, 1985).

Research work on pulses including cowpea was intensified with the initiation of All India Coordinated Project for the Improvement of Pulses in 1967. However, before the initiation of the Coordinated project, Indian Agricultural Research Institute (IARI) had released three varieties of cowpea viz. Pusa Barsati (vegetable type), Pusa Dofasali and Pusa Phalguni (both, dual purpose varieties). Several other varieties of cowpea were already under cultivation in different states of the country during sixties and the preceding decades. However, most of these varieties were grown for fodder or vegetable, and only a few for grain. The Publication and Information Directorate of CSIR listed about 58 cowpea varieties in its publication (The Wealth of India: Raw Materials, Vol. X during 1976). The most of the varieties listed in the publication were evolved either through introduction from foreign countries like Philippines, Kenya and USA; or through selection from the exotic and local land races. Only two varieties grown in Tamil Nadu for grain were developed through hybridization. No varieties were developed through mutation breeding approaches till seventies. IARI, New Delhi, in 1970 undertook the task of collection, evaluation, cataloguing and utilization of world genetic stock of pulses including cowpea. While Indian collection accounted for 716 germplasm accessions of cowpea, the exotic collection included 706 accessions representing 49 countries. The emphasis in cowpea breeding was for the development of varieties with compact plant type, non-trailing, early or medium maturity combined with high yield and resistance to yellow mosaic virus (YMV) disease. The first promising cowpea variety, C-152, released for general cultivation for grain production in 1976 was developed through selection from the cowpea cultivar P-42. In addition to IARI, cowpea improvement programmes have been going on at Indian Grassland Fodder Research Institute, Jhansi and at the Agricultural Universities of the states, particularly, in Haryana, Rajasthan, Gujarat, Maharashtra, Karnataka and Tamil Nadu.

Improvement of cowpea through induced mutagenesis

Mutation breeding research on cowpea in India was initiated in late sixties. Chemical mutagens were the first to be used for inducing mutations in cowpea. Sharma (1969a) at IARI, New Delhi and Narsinghani and Kumar (1969) at University of Udaipur carried out the studies using chemical mutagens. Sharma (1969b) reported very drastic mutations following treatments of cowpea variety Pusa Phalguni with different doses of dimethyl sulphate (DMS), ethylmethane sulphonate (EMS) and nitrosomethylurea (NMU). Mutations, besides affecting seed colour, had affected growth habit, flower colour, pod size, seed size and shape. This study also indicated the involvement of mutator system in seed coat colour. Narsinghani and Kumar (1969) studied the response of two cowpea varieties to the treatments with EMS (0.25%) and Methyl Methane Sulphonate (MMS) (0.025%) in the M_1 generation and observed reduction in germination and seedling height only in Pusa Barsati. Studies with physical mutagens were initiated only in eighties. Experiments for inducing mutations in cowpea cultivars using gamma radiation were carried out at the Regional Research Station, Bawal of the Haryana Agricultural University (Chowdhury, 1983); Marathwada Agricultural University, Parbhani (Borikar *et al.* 1983); and National Dairy Research Institute, Karnal (Vasudevan et al, 1984). Later, mutation experiments using gamma rays were conducted at other institutions including Tamil Nadu Agricultural University, Coimbatore; University of Agricultural Sciences, Bangalore; and Bhabha Atomic Research Centre, Trombay, Mumbai. Results of the various studies on induced mutations in cowpea conducted in India are summarized below.

Mutagenecity and Radio-sensitivity

Sharma (1969a) used chemical mutagens viz. DMS, EMS and NMU with the concentrations of 0.1-0.8%, 0.025-0.2% and 0.01-0.08%, respectively with an exposure period of 2-16 h for each treatment. On the basis of mutation frequency in the M_2 generation, the two mutagens DMS and EMS showed almost equal effectiveness (11.1 and 9.6%, respectively), while NMU was more effective causing 18.2% mutations. Sharma (1969b) observed mutations in several characters. Narsinghani and Kumar (1969) treated the presoaked seeds of cowpea cultivars, Red Seeded and Pusa Barsati with 0.25% EMS and 0.025% MMS for ten hours each and studied various parameters in the M_1 generation. Mahna *et al.* (1990) measured the effectiveness of Sodium Azide (SA) in alkaline medium on cowpea variety FS-68 and found that the treatment of pre-soaked seeds with different concentrations (10^{-6} - 10^{-3} M) of SA significantly reduced germination, root and shoot length, number of nodules and pods. Mahadevu *et al.* (1996) reported that plant height, number of leaves and branches reduced significantly in the M_1 generation by gamma rays treatment at the dose of 50 kR and above. There was significant reduction in number of flowers, pods, pod and seed weight at 40 kR in variety TVX-944 and at 50 kR in KBC-1. Pandey and Pawar (1998) used gamma rays dose of 20 kR to induce mutations in the cowpea variety V-130, and obtained a wide range of mutants in the M_2 generation. The selection of the dose (i.e. 20 kR) was based on the information generated by Brunner (1975) at the International Atomic Energy Agency, Vienna. Brunner (1975) had found 32.5 kR dose of gamma rays causing 50% reduction in shoot height and estimated 10-25 kR gamma rays dose to be useful for inducing mutation in cowpea. John (1999) however, found 30 kR gamma rays to be the efficient one for cowpea varieties Co-4 and C-152 and their hybrid on the basis of sterility and lethality studies. Our observation from the recently conducted experiments on an exotic cowpea, EC394763 treated with 25 kR showed that the germination was adversely affected at this dose, indicating the useful doses were lower to 25 kR. The various studies, however, have indicated the differential response of cowpea varieties to gamma rays treatment. It is, therefore, desirable to conduct experiments for radio-sensitivity in cases of new genotypes of cowpea.

CHLOROPHYLL MUTATIONS

Narsinghani and Kumar (1976) observed a large number of chlorophyll mutants in the M_2 generation of cowpea treated with 0.25% EMS and 0.025% MMS. *Albino, xantha, chlorina, striata* and *viridis* were observed in the EMS treatment, while only *albino, xantha* and *chlorina* mutants appeared in MMS treatment. John (1999) observed mutant *viridis* more frequently and *chlorina* and *xantha* in equal proportions followed by *albina* and *albo-viridis* following 30-40 kR gamma rays treatment of cowpea seeds of cultivars Co-4, C-152 and their hybrid. Pandey and Pawar (1998) also observed several chlorophyll mutants including *albino, xantha, chlorina* and *virescens* in the M_2 generation of cowpea variety V-130 treated with 20 kR gamma rays.

MORPHOLOGICAL MUTATIONS

Mutations affecting various plant parts and characters have been observed in cowpea following mutagenic treatments. Sharma (1969b) had observed very drastic mutations affecting several characters like growth habit, flower colour, pod size, spotting on pods, and colour, size and form of the seed in the M_2 generation following mutagenic treatments. Similarly, Pandey and Pawar (1998) reported a very wide range of mutations in the M_2 generation following 20 kR gamma irradiation of seeds of the cowpea variety V-130. The spectrum of mutations observed in this study is given in Table 1. Some of the mutants, especially those of seed coat colour, showed instability in the M_3 and subsequent generations. The appearance of a large number of mutants including some of the dominant ones in the M_2 generation (Table 2), and instability of mutant characters in the M_3 and subsequent generations confirmed the involvement of a mutator system or transposable element as reported by Sharma (1969b). Transposable elements can be activated under certain stress conditions like cell culture and gamma radiation (Walbot, 1988). Once the elements are activated, they can promote many kinds of mutations and chromosomal rearrangements. It is as if, transposable elements can amplify a small disturbance into a genetic earthquake. The distribution and organization of retro-transposable elements in cowpea genome has been reported (Galasso *et al.*, 1997). Mutations affecting plant parts and characters reported by various workers are summarized in Table 1.

Table 1. Spectrum of viable mutations in the M_2 in cowpea variety V-130 following 200 Gy* gamma rays treatment at Trombay.

Mutations affecting plant parts/ characteristics	Characteristics of Mutants	No. of mutants isolated
Growth habit	Erect	9
	Dwarf, bushy	3
Leaf shape	Altered leaf shape	5
Chlorophyll	Chlorina	6
	Virescent	8
Pod colour	Purple red	3
Seed colour	Altered seed colour	236
Seed size	Large	47
Seed shape	Altered seed shape	6
Maturity	Early	1
	Late	19
Sterility	Sterile	7

(* 200 Gy = 20 kR)

Growth habit

Sharma (1969a) had observed 'late giant' type mutants, characterized with large plant size, large leaves, peduncles, fruits, thick stems, large seeds, trailing growth habit and late maturity. Such mutations constituted about 11.3% of all the mutations recorded. Chowdhury (1983) obtained a bold seeded dwarf mutant of cowpea variety V-16 following seed irradiation with 40 kR gamma rays. Vasudevan *et al.* (1984) isolated two non-twining erect and two semi-erect mutants in the M_2 population of cowpea variety C-28 following irradiation of seeds with 20 kR gamma rays. Two mutants characterised with reduced internode length and fasciated stem were observed in M_2 generation of cowpea variety FS-68 treated with Sodium Azide (Mahna *et al.* 1990). A dwarf mutant with large seed size, isolated from the M_2 generation and grown in the M_3 and subsequent generations was found true breeding only in *rabi* season (Pandey and Pawar, 2000).

Table 2. Spectrum of seed colour mutations in M_2 of cowpea cultivar V-130 following 200 Gy gamma rays treatment at Trombay

Seed colour	No. of mutants
Stippled black	1
Grey	2
Brown	5
Pinkish brown	5
Yellowish brown	3
Light pink	5
Light yellow	5
Light green	11
Bright white	8
Partial black	38
Partial grey	22
Partial purple	27
Partial brown	43
Partial chocolate	36
Partial pink	9
Partial yellow	5
Partial green	5
Ivory white	5
Dotted brown	1
Total	**236**

('Partial' refers to a large coloured area around hilum)

Leaf, flower and seed characters

Borikar *et al.* (1983) isolated three small leaf mutants in cowpea variety 2-1 and four broad leaf mutants in variety C-152 following gamma rays treatment. Delayed flowering by 15 days was observed in a mutant plant obtained from 10^{-4} M of Sodium Azide treatment (Mahna *et al.*, 1990). Chowdhury (1983) obtained a bold seeded mutant with 40% increase in weight following 40 kR gamma rays treatment of cv. V-16. A brown seeded mutant of cowpea was isolated from the M_2 generation of the white seeded variety Pusa Phalguni following seed treatment with 0.8% DMS for 2 h (Kharkwal *et al.* 1988). The mutant however, showed genetic instability in subsequent generations because of probable induced mutator system. A number of stable lines with altered seed colour as well as improved agronomic characters were obtained from the later generations in which the mutator activity was lost. Pandey and Pawar (1998) isolated several seed coat colour mutants from the M_2 generation after the treatment of a white seeded variety V-130 with 20 kR gamma rays. The seed colour mutants varied from light brown, dark brown, light green, light yellow, bright white, creamy white, light grey to dark grey to stippled black (Table 2). The dark grey and stippled black seed colour mutants showed instability in subsequent generations and segregated into various seed colours indicating the involvement of transposable elements.

Disease resistance

An induced mutant tolerant to seed borne mosaic virus disease was identified (Choulwar and Borikar, 1986a). A mutant variety V-240 developed from Pusa Phalguni following DMS (0.8%) treatment has shown resistance to all the major fungal, bacterial and viral diseases (Kharkwal *et al.* 1988). Mahadevu *et al.* (1998) isolated a *Pythium* rot resistant mutant from the M_2 population of cowpea variety KBC-1 treated with gamma rays. When screened for seedling mortality in the M_3 generation, the mutant showed resistance with less than 5% mortality against almost 87% shown by the parent.

Nodulation

Maherchandani and Rana (1977) obtained ten lines superior in nodulation, from the M_4 population of C-152 treated with gamma radiation. A high nodulating mutant of cowpea was obtained following treatment with 10^{-5} M Sodium Azide (Mahna *et al.* 1990).

Yield and Yield components

Induced genetic variability for number of branches, pods per plant, pod length and seeds per plant was observed in M_3 generation following 20 and 30 kR gamma ray treatments (Choulwar and Borikar, 1986b). Phenotypic and genotypic coefficients of variation, heritability estimates and genetic advance for seed yield per plant were higher in lines derived from 10 kR treatment. Induced variability in cowpea varieties KBC-1 and TVX-944-02E following 10-50 kR gamma rays treatment was found maximum for plant height, number of leaves, pod weight, shoot biomass and seed yield in M_3 generation (Mahadevu *et al.* 1998). High yielding mutant lines viz. V-16, V-240 and V-38 were selected from the stable lines generated by 0.8% DMS treatment of the variety Pusa Phalguni (Sharma and Kharkwal, 1983, Kharkwal, *et al.*, 1988). The mutants V-16 and V-240, both with brown seed coat and V-38 with bi-colour seed coat characterized with early maturity, non-trailing growth habit, synchronous flowering and better quality of pods and grains were released as varieties. Mahna *et al.* (1990) found a high yielding mutant with increased number of flowers, pods and seeds in cowpea variety FS-68 following treatment with 10^{-5} M of Sodium Azide. Radiation induced promising mutants, TCM77-4, TCM13-5, and TCM121-8 were obtained by Pandey and Pawar (2000). The mutant TCM77-4, with its reduced plant height, bushy growth and large seed size was envisaged to be the most suitable for *rabi* cultivation, especially in rice fallows. The mutants TCM13-5 and TCM121-8 with their respective large seed size and pod number show great promise for higher yield.

Varieties

Cowpea varieties developed through induced mutation and released for commercial cultivation in India are given in table 3, while varieties Co. 5 and Cowpea -88 were developed for the fodder, other varieties were of grain types. A brief description of these varieties is given in Table 3.

Table 3. Cowpea varieties developed through induced mutations in India

Variety	Year of release	Institution	Mutagenic treatment and Parental variety
V-16 (Amba)	1981	IARI, New Delhi, India	DMS, Pusa Phalguni
V-37 (Shreshtha)	1981	IARI, New Delhi, India	DMS, Pusa Phalguni
V-38 (Swarna)	1981	IARI, New Delhi, India	DMS, Pusa Phalguni
V-240	1984	IARI, New Delhi, India	DMS, Pusa Phalguni
Co. 5 (forage cowpea)	1986	TNAU, Coimbatore, India	30 kR γ- rays, Co. 1
Cowpea-88	1990	PAU, Ludhiana, India	F_1 seed irradiation (Cowpea-74 X virus-resistant strain No. H-2)

V-16 (Amba)

It is a brown seeded mutant variety developed for higher grain yield by treating the seeds of white seeded variety Pusa Phalguni with 0.8% of DMS for 2 hours and isolating the mutant. Identified for release in 1981, V-16 was released for cultivation in 1984 by Govt. of India for all cowpea growing areas of the country. It was described as the highest yielding variety in India and was resistant to fungal and bacterial diseases.

V-37

This variety was also evolved through induced mutagenesis of Pusa Phalguni by DMS treatment. It was high yielding with luxuriant vegetative growth, and was, therefore, suitable for green fodder also.

V-38 (Swarna)

Developed from the DMS treatment of Pusa Phalguni, the variety V-38 is early maturing, non-trailing with synchronous flowering, better quality pods and grains. It is good for grain as well as for vegetable purposes. Seeds are moderately small and roundish in shape. The seed is bi-coloured, fawn around the hilum and red on the remaining part of the testa. V-38 was almost immune to most of the diseases.

V-240

Obtained through the mutagenic treatment of Pusa Phalguni with DMS, V-240 is a brown seeded mutant released for cultivation in 1984 for all cowpea- growing regions of the country. It was high yielding, medium late and resistant to all major fungal, bacterial and viral diseases.

Co - 5

The variety Co - 5 was developed for forage at the Tamil Nadu Agricultural University, Coimbatore, by irradiating the seeds of cowpea cultivar Co -1 with 30 kR gamma rays. It was described to be more nutritive giving 16% higher yield than the parent. It is compatible for inter-cropping with fodder cereals.

Cowpea-88

Released for cultivation in Punjab in 1990, Cowpea-88 was developed by irradiating the F_1 - seed of the standard variety Cowpea-74 and a virus resistant strain no. H-2. It is a dual purpose variety which gives higher green fodder as well as higher grain yield than Cowpea-74. It is characterized with large seed size and chocolate brown seed colour. Cowpea-88 is highly resistant to yellow mosaic virus and anthracnose diseases.

Future prospects

The cultivation of cowpea is concentrated in arid and semi-arid areas of Rajasthan, Gujarat and Karnataka; it is desirable to develop extra-early varieties with reasonable yield. Extra-early varieties (60 days maturity) will specially suit to such a conditions. Further, there is a need for defect elimination in local photo-period sensitive varieties. Development of varieties with tolerance to drought and high nitrogen fixing ability can maximize the grain and fodder production in marginal drought prone areas. There is need for developing varieties with combined resistance to major diseases. Future varieties with erect, compact and terminally bearing growth habit are likely to change the face of cowpea in India. White large seeded varieties are likely to fetch better market price and will prove to be remunerative to farmers. All these can certainly be achieved by conventional and mutation breeding approaches.

Extensive programmes for cowpea improvement have been going on at the IITA, Nigeria; which has become the major source of cowpea germplasm and information on its breeding. The cowpea germplasm exhibits large genetic diversity for plant type, growth habit, maturity and other plant characteristics, and thus provides opportunity to breeders for introduction, selection and hybridization. Induced mutation, a tool for creating additional variability and for selecting desired types in relatively shorter time span, has been used to improve cowpea. Mutation breeding approaches are considered more effective in incorporating recessive traits and could be used to convert some of the tropical cowpea accessions in the world germplasm collection into early flowering types that would be more useful in breeding programmes in sub-tropical zones (Hall *et al.* 1997). Mutation breeding approaches may effectively supplement the conventional breeding programmes, particu-larly in India.

REFERENCES

Borikar, S.T., Hudge, V.S. and Salunke M R. 1983. Induced leaf mutants in cowpea. *Tropical Grain Legume Bulletin* **28**: 8-10.

Brunner H. 1975. Radiosensitivity of grain legume species. *Mutation Breeding Newsletter* **5**: 7-9.

Choulwar S B and Borikar S T. 1986a. Induced tolerance in seed borne mosaic virus of cowpea. *Journal of Maharashtra Agricultural Universities* **11**: 244-245.

Choulwar S B and Borikar S T. 1986b. Induced genetic variability in cowpea. *Journal of Maharashtra Agricultural Universities* **11**: 255-256.

Chowdhury R K. 1983. A bold seeded dwarf mutant of cowpea. *Tropical Grain Legume Bulletin* **27**: 20-22.

Galasso L, Harrison GE, Pignone D, Brandes A and Heslop-Harrison J S 1997. The distribution and organisation of Ty1-copia-like retro-transposable elements in the genome of *Vigna unguiculata*

(L.) Walp. (Cowpea) and its relatives. *Annals of Botany* **80**: 327-333.

Hall A E, Singh B B and Ehlers J D 1997 Cowpea Breeding. *Plant Breeding Reviews* **15**: 215-274

Jeswani L M and Baldev B (Ed.). 1990. *Advances in Pulse Production Technology*, 105-109 pp. I.C.A.R., New Delhi.

John S A 1999. Mutation frequency and chlorophyll mutations in parents and hybrid of cowpea following gamma irradiation. *Indian Journal of Genetics and Plant Breeding* **59**: 353-361.

Kharkwal M C, Jain H K and Sharma B. 1988. Induced mutation for improvement of chickpea, lentil, pea and cowpea. (in) *Proceedings of Workshop, Pullman, 1-5 July, 1986, on Improvement of grain legume production using induced mutations, 89-109 pp* .IAEA, Vienna.

Mahadevu P, Thimmaiah S K and Kulkarani R S. 1996. Induced variability in gamma-irradiated seeds of cowpea (*Vigna unguiculata* (L.) Walp) for growth, nodulation and yield. (in) *DAE-BRNS-Symposium on Nuclear techniques in increasing crop and animal productivity, 69 pp.* B.A.R.C., Mumbai.

Mahadevu P, Thimmaiah S K and Kulkarni R S. 1998. Induced variability and character association in gamma irradiated seeds of cowpea (*Vigna unguiculata* (L.) Walp.). (in) *DAE Symposium on Induced mutations and molecular techniques in improving crop productivity and quality. 9 pp.* Chennai,

Mahadevu P, Thimmaiah S K and Kulkarani R S. 1998. A Pythium rot resistant mutant of cowpea (*Vigna unguiculata* (L.) Walp.) cv. KBC-1. (in) DAE-Symposium on Induced mutations and molecular techniques in improving crop productivity and quality. Oral presentation-50. Chennai.

Maherchandani N and Rana O P S. 1977. Gamma radiation induced natural variability for nodulation in legumes. *Journal of Nuclear Agriculture and Biology* **6**: 75-77.

Mahna S K, Bhargava A and Mohan L. 1990. Alkaline azide mutagenicity in cowpea. *Mutation Breed. Newsletter* **36**: 6-7.

Narsinghani V G and Kumar S. 1969. Induction of mutants in *Vigna sinensis* Linn. (in) Proceedings of Symposium on Radiation and radiomimetic substances in mutation breeding, Sept. 26-29, 371-374 pp. Bombay.

Narsinghani V G. and Kumar S. 1976. Mutation studies in cowpea. *Indian Journal of Agricultural Sciences* **46**: 61-64.

Pandey R N and Pawar S E. 1998. Induced mutations for improving plant type, quality and yield in cowpea. (in) DAE-Symposium on Induced mutations and molecular techniques in improving crop productivity and quality, 2 pp. Chennai.

Pandey R N and Pawar S E. 2000. Radiation induced promising mutants in cowpea. *(in) Proceedings. Of DAE-BRNS Symposium, Dec. 6-8, 2000, 166-169 pp.* B.A.R.C., Mumbai.

Sharma B. 1969a. Chemically induced mutations in cowpea (*Vigna sinensis* L., Savi). *Current Science* **38**: 520.

Sharma B. 1969b Induction of a wide range mutator in Vigna sinensis. (in) Proceedings of Symposium on Radiation and radiomimetic sustances in mutation breeding. Sept 26-29. 13-21 pp. B.A.R.C., Bombay.

Sharma B and Kharkwal M C. 1983. Mutation breeding of lentil, cowpea and chickpea. *Mutation Breeding Newsletter* **21**: 5-6.

Singh B B and N'tare B R. 1985. Development of improved cowpea varieties in Africa. *(in) Cowpea Research, Production and Utilization, Sigh S R and Rachie K O (Eds), 105-116 pp.* John Wiley and Sons, England.

Singh B B and Sharma B. 1996. Restructuring cowpea for higher yield. *Indian Journal of Genetics and Plant Breeding* **56**: 389-405.

Singh B.B., Chambliss O.L. and Sharma B. 1997. Recent advances in cowpea breeding. *(in) Advances in Cowpea Research, 30-49 pp. Singh BB, Mohan Raj D R, Dashiell K E and Jackai L E N (Eds).* Co-publication of International Institute of Tropical Agriculture (IITA) and Japan International Research Centre for Agricultural Sciences (JIRCAS). IITA, Ibadan, Nigeria.

Singhal V (Ed). 1999. Miscellaneous pulses: Lobia (Cowpea). (in) *Indian Agriculture, 129 pp.* Indian Economic Data Research Centre, New Delhi.

Smartt J.1985. Evolution of grain legumes III. Pulses in the genus *Vigna. Experimental Agriculture* **21**: 87-100.

Vasudevan K, Shanker K. and Dua R P. 1984. Gamma ray induced erect and semi-erect mutants in *Vigna unguiculata* (L.) Walp. Mutation Breeding Newsletter **24**: 4.

Walbot V. 1988. Reactivation of the mutator transposable element system following gamma irradiation of seed. *Molecular and General Genetics* **212**: 259-264.

CHAPTER - 7

BREEDING COWPEA FOR ABIOTIC STRESSES TOLERANCE

D. Kumar

Central Arid Zone Research Institute, Jodhpur

ABSTRACT

Cowpea, is typically characterized with drought tolerance potential through avoidance phenomenon contributed by certain adaptive, morpho-physiological traits: dynamic root system, broad drooping leaves, stomatal sensitivity, high xylem pressure potential, high leaf area at the end of flowering etc. However, being high in photosynthetic activity, cowpea is poor in carbohydrate partitioning towards the sinks. Looking at its growing habitats, it is argued that selection for earlyness, followed by improvement in avoidance may be prioritized in cowpea, as earlyness will lead to high yield without serious impacts on depeletion of soil moisture and rising temperature also. The selections may be made in wider inter row spacings in targeted regions only. Effective screening procedures using appropriate selection based markers in situations representing warm and dry climate is of great importance. Extension of leaf area, specific leaf weight, phenomenal dry matter partitioning and number of mature pods are the sensible traits could be used for selection of lines representing drought tolerance. For hybridization, lines having avoidance and those with escape mechanisms may be hybridized for combining the genes for both mechanisms of tolerance. Lines having maximum absolute yield and those with lesser proportion of reduction in dry than on optimum conditions may also be isolated separately for furthering the yield improvement on drought prone situations.

Salinity influences cowpea through osmotic, ionic and toxic effects, causing serious reduction in grain yield. Cowpea is however, medium tolerant to salinity with a threshold salinity level of ECe 4.9 dSm^{-1}, however it is regarded as quite sensitive at the vegetative stage and the tolerance increases with advancement in growth. Thus, fortunately it can tolerate salinity at the reproductive stage to which majority of the crops are referred to as sensitive. In terms of soil alkalinity, cowpea is understood to be more sensitive to salinity. Breeding strategy for salt tolerance involves effective screening of germplasm collected from salt habitats at critical growth stages using at least three levels of salinity i.e., ECe 6, 9 and 12 dSm^{-1}. The screening may be done in micro-plots with sinthetic saline water.

Heat effects are confounded with drought effects. Generally high night temperature and longer photo periods (>12 h) may prove very detrimental for cowpea. Not the whole plant but individual cells may get strains from heat shocks. However, highest injury has been observed at the pod filling stage due to increased pollen sterility. Tolerance to heat is achieved by two way of avoidance and tolerance phenomena. The former is achieved through: transpirational cooling, reflection of solar radiation through wax deposition or development of hairs. Tolerance is, whereas achieved through, biomembrane saturation and its protection through synthesis of certain enzymes like, peroxidase and superoxide dismutase. Heritability in dry and hot climate is low hence, transfer of heat tolerance may pose some problem, even thought, heat tolerance is governed through few major genes in cowpea. Breeding for heat tolerance may involve effective screening through Electrolyte Leakage method and selecting fertile lines at high embient temperature. Heat tolerance may involve excessive broad leaves, hairyness and deposition of wax on them, medium early to early maturity.

Lines tolerant to drought, heat and salt stresses have not been bred so far in any crop including cowpea, those reported showing tolerance to one or more than one stresses at a time are the consequent of their evaluation in optimum situations but sowing better perfor-mance in such situations also. Breeding for tolerance to these stresses is, therefore, still in take-off stage.

INTRODUCTION

For field crops, stress can be defined, a factor which, irrespective of its nature, source and origin intervenes with the growth of plants from germination to maturity. Of a number of stresses, abiotic stresses are the ones greatly accentuated in resource constraint arid situations producing adverse impacts of varied and unpredicted nature and extent. The same are induced due probably to water scarcity (drought or desiccation stresses), or excess of water (waterlogging); increased salt content in the soil or in the irrigation waters (salinity/alkalinity); increased chemicals concentration (chemical stress such as metal or pH stresses); increased level of pollutants (pollution stresses), increased day/night lengths (photoperiodic stresses); supra-optimal (high temperature stresses) and sub-optimal temperature (freezing, frost and cold stresses) and still many more at the soil surface and crop stand levels. These stresses exhibit compound effects in practical sense when more than one stresses occur at the same time interval, as for example, oxidative, photo and thermo-stresses. These stresses have been causing many social, economic and cultural problems, particularly in developing world on marginal and submarginal lands. Abiotic stresses are known for exerting maximum dimensions and appear to be the sole causes decreasing the sustainability and productivity of field crops. For instance, according to Dulal (1976) only about 10% of the total arable lands world over may be considered as free from stresses. Relative damage of the abiotic stresses however, depend on the crop, environment and the obligatory secondary or tertiary stresses. The stresses caused due to drought, salt and heat are the most prevalent ones world over. Statistically, of 1474 m ha total arable lands world over, unirrigated, rainfed lands (water scarcity) are 1247 m ha (84%). Of 6.0 billion ha of arid and semi-arid lands, 1.0 billion ha are affected by excess of salts. Similarly of 0.27 billion ha of irrigated lands, one-third are believed to be affected by salts (Eckholm, 1975). Whole of the arable rainfed-lands in arid and semi-arid regions (low rainfalls) are subjected to heat stress causing considerable damage to the crops.

Cowpea anciently known as *lobia,* is traditionally the crop of arid and semi-arid regions adequately and efficiently acclimatized to large extent of rainfall and soil variations. The drooping and broad leaves, help in reducing soil temperature, therefore, conserving soil moisture and reducing soil movement to some extent. Initial fast growth of cowpea enables it to early suppression of weeds all around, consequenting in reduced canopy-weed competition and early adaptation to soil adversity. Besides, this important legume has established its better popularity than the grain legumes due to its economic importance, nutritional and medicinal merits attached to it. Cowpea, therefore, due to adaptive and remunerative values is characterized the most important pulse of marginal and sub-marginal arid farm lands. Ability of this crop to be grown in all the three seasons of tropical Indian sub-continent is, gaining all round ground, even in non-traditional areas.

Cowpea is the crop of dry and warm habitats and flourishes very well on infertile soils of arid and semi-arid zones world over. Drought and heat independently or through their combined strains can substantially reduce cowpea productivety (Nielson and Hall 1985b, Hall and Patel 1987, Warrang and Hall 1983, 1984). Salinity may also prove quite detrimental for cowpea due to less precipitation in the soil during rainy season, being inadequate to leach the salts down, accumulated due to secondary salinization process. Thus, drought, heat and salt stresses are the major abiotic constraints, which may prove as main yield determinants due to their primary effects (osmotic, ionic and transpirational stresses) and secondary/tertiary common path ways like dehydration strain to the cowpea plants.

Cowpea in general, exhibits tolerance to heat and drought stresses but may only show moderate tolerance potential towards varied concentrations and quality of salts. The simple solution to these stresses is, providing adequate good quality water to cowpea which is scarce and its transport may prove the most costly management input under rainfed conditions. Thus, identification of cowpea genotypes which can tolerate these stresses is strongly desired and may prove sustained management strategy. In view of it, efforts through this compilation have been made to provide conclusive and comprehensive information on responses, screening measures, selection arids, tolerant genotypes and breeding strategies for cowpea improvement in respect of these important abiotic constraints, addressed in reference to cowpea.

DROUGHT

Drought induced through water scarcity is a natural cycle and is subjected to many places in the agricultural world. However, the intensity and duration may vary considerably from one climatic zone to next. Similarly, the magnitude of adversity may also vary and fluctuate with the crops, genotypes and the growth stages. Losses incurred from intermittent drought cycles may be direct emanating from the reduced crop yields, and indirect emerging from the abandment of crop lands, non-plantation of crops, change in land use pattern, reduced employment and huge revenue losses. The primary effects of droughts emanates from reduced agricultural production, which may envelope whole country when State Governments have to spend huge exchequer towards drought relief in many ways. In India, of total geographical area (329 m ha), about 143 m ha is arable annually, however, the rainfed area (95 m ha) constitutes 68% of the net sown area. However, even if fullest irrigation potential is exploited to manage drought, almost 113 m ha can hardly be brought under irrigated agriculture by 2025 AD, leaving still 47% area depending on rains for their cultivation. Thus, needed reliance on remedial measures for enhancing production of crop plants from rainfed lands, are required to be continued.

Cowpea is an important arid legume largely grown as rainfed crop world over particularly, in harsher areas of Sahalian and Sudanian zones of Asia and the regions of Middle -East and South-East Asia. Inspite of inbuilt drought tolerance potential, it suffers heavy yield and biomass losses in response to highly fluctuating environmental vagaries. In view to accelerate improvement in production potential, breeding for drought resistance seems to be the realistic approach. The same underlines the importance of response to drought, drought tolerance mechanisms, genetics of drought tolerance and suitable methodologies which are underlined below:

(i) Response

Morphological: The response of plants to drought may be highly variable, complex and greatly fluctuate and accentuate by a variety of interacting factors. In general, the exposure of cowpea plants to moisture stress is known to depress and delay the germination of seeds and seedlings emergence, reduction in water uptake, mobilization of dry matter reserves even at the early growth stages, ultimately affecting growth and production. For instance, water shortage reduced stem length, number of internodes, leaves and shoot dry weight in cowpea in Cairo situations (Khalil and Mandurah, 1989).

In cowpea, increased level of water scarcity have been observed to reduce the plant stature, leaf area and reduced grain yield by reducing the number of pods m^2 and seed size (Turk *et al.*, 1980). In Ibadan situations, also, grain yield was found to decrease when soil water stress was induced at three growth stages in cv. Adzuki (erect), Ife Brown (semi-erect) and New Era (spreading). Furthermore, floral abortion increased with a decrease in matric potential (Babalola, 1980). Reduction in grain yield, in response to water stress varied with the determinate (C 152) and indeterminate (C 779) cultivars and was more in the latter in Delhi situations (Chaturvedi *et al.*, 1980). Alongwith grain yield, the proportion of DM also varied in response to drought. For instance, there was greater proportion of shoot DM in reproductive parts during early pod filling stages; and was greater with moderate drought, surprisingly the proportion of the same remained unaffected with severe drought in comparison to irrigated conditions at California (Grantz and Hall, 1982). The DM production in cowpea under limited water conditions was linearly related to crop water use but relatively insensitive to the time of water stress (Shouse *et al.*, 1982). Not only dry soil but dry air also affected the growth of cowpea plants. For instance, dry air caused significant reduction in whole plant biomass despite the fact that plant water status was not different to humid air at Germeny (Nagarajah and Schuleze, 1983). Dry soil decreased biomass and growth but leaf growth continued at which stomata began to close down. Thus, cowpea stomata appear more sensitive than leaf growth. Relative leaf expansion rate started to decline when 60% (humid air) or 65% (dry air) of plants available were depleted. However specific leaf weight increased with increasing level of drought.

Drought is understood to induce distinct effects on earlyness also. The earlyness of cowpea was increased with moderate drought but decreased with more severe drought. In general, however, severe drought tended to shorten the reproductive phase and extended the vegetative phase. In a significant finding made by Turk and Hall (1980), it was observed that a multiple correlation of seed yield with specific leaf weight and degree days (above 35°C) at flowering stage, accounted for most of the variations in seed yield during progress drought (Turk and Hall, 1980).

Growth stage: Grain yields in most crops are understood to be susceptible to water defecits at flowering (Jordan *et al.*, 1983), when seed number is established. Stress at flowering being quite common in cowpea grown in arid areas, may lead to losses of capital and food, therefore, can be serious for resource-scare cowpea farmers.

Seed yield was decreased by water deficit occurring at flowering and pod filling stages and was linearly related to an integrated value of predawn LWP from bud formation to pod fill stage (Shouse, 1979). Study at Riverside revealed that cowpea maintained seed yield when subjected to drought during vegetative stage when subsequent irrigation intervals provided were not sufficient. Similarly, drought stress when created at the vegetative stage did not significantly reduce the seed yield of cowpea cv. TVU-1954-OIE at USA. The WUE of plants also significantly increased when the plants were water stressed at vegetative stage. Drought stress, however when induced at flowering, reduced the number of pods by about 50%. At Ibadan, maximum seed yield was obtained when water stress was induced at the flowering-podding stage in cv. Ife Brown (Babalola, 1980). Convincely, the yield losses in field crops may be ascribed to inadequate water availability at the critical sensitive growth stages (Boyer, 1996). Crops including cowpea, therefore, differ in their susceptibility of growth stages to water stress environment.

The WS subjected at the growth stages of cowpea, significantly affected the contents of protein, cysteine, threonine, valine methionine and isoleucine amino acids. WS at flowering

Most Promising cowpea genotypes

Most Promising cowpea genotypes

Sporadic pods, lack of sink in susceptible genotype

Drought tolerant strain with profuse bearing during
Kharif, 2002

and pod filling stage increased the free amino acids pool and at the same time, inhibited the incorporation of the amino acids into the protein chain, thus, lowering the protein amino acids fraction simultaneously (Labanauskas *et al.*, 1981).

Physiological: Water stress exposed cowpea plants immediately lower down their relative water content (RWC) of the leaves, leaf water potential (ψ l), osmotic potential (ψ); and turgor pressure (TP) may also decrease unless the plants have finally adapted to drought by opening their stomata again and improving the water potential. In cowpea, stomata appear more sensitive to water stress than the leaf growth. Thus, a common response to water stress is the closure of stomata, which reduces the fluxes of both CO_2 and water vapors. Alternatively, stomata may remain open while turgor is maintained through osmotic adjustment. The latter response should result in a greater C gain. On withheld of irrigation, the cowpea plants adjusted slowly but maintained a high WP by stomata closure.

The stomata generally close down in dry air and water status remained lower in dry than in humid air due to smaller leaf area and the lower leaf conductance. As usual, a decrease in OP of sub-strata produced in seedlings of cowpea (cv. Pusa Barsati) resulted in decreased RWC, LWP and increased the rate of respiration, proline content and the activities of IAA oxidase, ascorbic acid oxidase, peroxidase and glycolate oxides (Mukherjee and Choudhuri, 1983), helping cowpea plants to adjust osmotically to decreasing WP. Water stress when induced in laboratory through PEG-6000 at Burdwan, it increased the water saturation deficit and proline and protein content, and decreased LWP, and chlorophyll (Biswas and Choudhuri, 1986).

Microbiology: Water stress noteably affects the nodulation process and other enzymatic activities decreasing therefore, growth process considerably. Maximum nodulation of 175 mg/ plant was observed at flowering stage in well watered conditions than droughted plants. Following flowering stage there was no difference in nodulation with water regimes. Nitrogenese activity decreased in stressed than unstressed plants. According to an estimate, during dry regime, considerably lower nitrogenase activity of 5.3 and 4.5 μ mole ethylene/plant/hour at pod filling and pod yellowing stages resp., than 22.5 and 18.8 μ mole ethylene/plant/hour in cowpea in irrigated conditions at the respective stages was observed (Robert *et al.*, 1981). Nitrogenase activity completely ceased in cowpea with decease in LWP from –0.4 to –0.9 MPa. With increasing WS, the acetylene reduction activity (ARA) declined shaply in cowpea than in groundnuts. Even with mild WS (LWP of 0.2 MPa) nodule FW declined by 50% in cowpea (Venkateswarlu *et al.*, 1989).

Biochemical: Plants including cowpea when exposed to different levels of water stress react biochemically and the consequent synthesis and accumulation of certain organic substances help them to osmotically adapt by maintaining internal water status and the turgor pressure. Plants, for instance, show synthesis and accumulation of certain metabolites like proline, glycine, betaine etc.; following stresses. Sugar or sucrose accumulation during drought may also help plants by preventing coagulation and desiccation of the cells and contributing to drought tolerance (Turk *et al.*, 1980). Moreover, plants may also show thickening of protoplasm while stimulating lipids leading to elimination of vacuoles and drought tolerance.

Some of the stresses to which plant species are exposed initiate the synthesis of certain metabolites which may prove useful and toxic as well at the cellular and whole plant level. The same result from the biochemical and physiological reactions taking place within plant cell. Certain toxic oxygen elements directly affect the normal stature of cell ingradients, like protein degradation, DNA strand breaking and pigment degradation may also take place. Intigrated activities of the anti oxident defence system protect plant cells from the deleterious effects of active oxygen species produced at the elevated rates when plants are exposed to environmen-tal-stress conditions. Water stress decreases shoot total and protein N contents but increased the soluble N contents in cowpea at Cairo situations (Khalil and Mandurah, 1989).

(ii) Adaptation and Tolerance

Checking the susceptibility or acclimation are the responses plants exercise when exposed to varied levels of water stress. The inherent genetic plasticity leading to acquired mechanism of tolerance in response to stress is important attribute involving positive genetic selection. Some adjusting or adaptive traits (i.e. flowering and maturity) in response to stresses have been observed to provide with needed high plasticity, contributing to immediate adaptation towards the existing hostile situations. Generally plant tolerance to soil moisture deficit is achieved in many ways: proportional and sequential growth retardation, diversion of carbohydrates to essential parts, physiological potentiality and synthesis and accumulation of biochemical substances. Drought tolerance mechanisms are not however, so simple and confining to these features only. The further improvement comes from cell membrane and individual cell to whole plant, involving synthesis of specific proteins, osmolytes and other complex pathways.

Cowpea is known for a dynamic root system which is tuned to soil moisture status and regulates the limited water available to the plants due to its effective distribution in the soil zone. Drought adapted cowpeas are known to have larger and well adapted root system (Itani *et al.*, 1992). Cowpea adapt to drought by exhibiting extreme drought avoidance phenomenon in the presence of soil water deficit or during high evaporative demands with xylem pressure potential remaining above -18 bar in Riverside situations (Turk and Hall 1980). Earlyness, increase in leaf area and high leaf area index at the end of flowering were related to drought tolerance in cowpea (Turk and Hall *et al.*, 1980). Drought tolerance in cowpea at Texas was achieved through high LWP and higher acetylene reductase thus, resistance was marked by more nitrogen fixation (Walker and Miller 1982). Adaptation of cowpea in harsh environment may be generally achieved through early partitioning of carbohydrates (Grantz and Hall, 1982). Higher water content in detached leaf in cowpea gave tolerance to drought. The extent of osmoregulation capacity has been shown to be positively related to growth and yield in pulses (Morgan *et al.*, 1991).

Osmotic and passive water uptake abilities in cowpea may be lower than some of other pulses, but cowpea had a deep root system and roots further developed vertically when plants were exposed gradually to water stress. Moreover, cowpea has low water uptake ability hence, suited nicely to water stress situation.

Resistant genotypes in cowpea were capable of maintaining high LWP and biomass production during water stress. In addition, the effect of water stress on N_2 fixation was far greater than the genotypes alone selected for drought tolerance. Thus, drought resistance in cowpea is understood to be through efficient nitrogen fixation (Walker and Miller 1986).

Ten crop species were evaluated for their relative drought tolerance at the seedling stage. Based on percentage mortality rate at various intervals and days taken to 100% mortality, soybean appeared the most drought susceptible and cowpea the most drought tolerant one (Singh *et al.*, 1999). The overall ranking of the crops in the increasing order of drought tolerance was : soybean (*Glycine max*) <black gram (*V. mungo*) <green gram (*V. radiata*) <groundnut (*Arachis hypogea*) <maize (*Zea mays*) <sorghum (*Sorghum bicolor*) <pearl millet (*Pennisetum glaucum*) <bambora nut (*V. subterranean*) <lablab bean (*Lablab purpureus*) <cowpea (*V. unguiculata*).

Relative growth rate differed with tolerance potential at the genotypic level. Susceptible genotypes showed the closure of stomata very rapidly at lower leaf predawn water potential (psiw) of -0.6 MPa than the tolerance ones -0.9 MPa. Net assimilation rate was closely coorelated with stomatal conductance. Tolerance was associated with better tissue water retention capacity, moreover leaf dehydration rate was also slow. These results confirmed by Electrolyte Leakage Test and OP measurements, revealed that in *P. vulgaris* higher membrane resistance and O. A. were involved in drought tolerance (Franca *et al.*, 2000).

Drought limits the agricultural production by preventing the crop plants from expressing their full genetic potential. In achieving so, plants adapat and tolerate the drought by many ways involving many complex process at different levels. Preeisouly however, three mechanisms: escape (early partitioning and maturity), avoidance (morphological adaptations) and tolerance (physiological and biochemical adaptations) confer resistance to moisture deficit situations. Conclusively it can be resolved that cowpea is known for inbuilt tolerance potential towards drought and it is achieved by the following mechanisms:

1. Initial faster growth than other legumes confirming operation of drought avoidance phenomenon (lawn, 1983).
2. Deep and fast penetrating root system.
3. Higher number of stomata on lower than upper surface and higher stomatal sensitivity compared to leaf growth may regulate water as per need (Itani, 1992).
4. It is known as dehydration avoider due to stomatal sensitivity.
5. Translocation of larger proportion of dry matter towards roots than towards pods at early stage (Watanable *et al.*, 1997).
6. Cowpea is known for greater degree of stability in phosholipase D enzyme, which is responsible for lipid degradation, thus contributing towards its tolerance to water stress (El. Maarouf *et al.*, 1999).

(iii) Indices of tolerance

Increase in production potential of cowpea to date has consequented mainly from evaluation of germplasm for high yield only. However, evaluation in drought prone situations based on probable drought indices are still in take off stage. Generally, drought tolerance mechanisms being complex and variable are by and large supposed to be associated with thickness of cuticle, stomatal conductance, root growth and its extension, hormonal and enzymatic activities, remobilization of carbohydrates, O.A. and antioxidant productions. However, specific and convenient quantitative traits referred to cowpea only, which can help screening of noble germplasm lines for drought tolerance are listed below:

1. In rainfed (277.5 mm) situations of Burkino Faso cowpea varieties with high plant dry weight yielded the best particularly, in drought situations. Plants with vigorous growth appeared to have an inherent ability to recover from drought (Aggarwal, 1987).
2. Number of pods/plant, 100-seed weight and number of seeds/pod were the primary yield components in cowpea for drought situations. Subtle indirect effects of vigour index and number of pedeuncles/plant on

grain yield were more important than their direct effects in Nigerian conditions (Obesesan, 1985).

3. Loss in fresh leaf weight (ca 7%) at peak frequency of dehydration compared to original fresh weight of cowpea leaf during fixed incubation period in Italy situations was suggested as a rapid appraisal method (Shadeya *et al.*, 2000).
4. A direct correlation was achieved with respect to chlorophyll stability index (CSI) and yield reduction in rainfed situations in cowpea at Jhansi. Grain yield, RWC and CSI showed significant positive associations with each other (Yadav and Patil, 1984).
5. Carbon isotope discrimination (DELTA) has been proved as a selection criterion for improving the adaptation to water limited environment because it provides a measure of seasoned tranpiration.
6. High leaf water potential and acetylene reduction in cowpea indicating more N_2-fixation (Walker and Miller, 1982).
7. Detached leaf of cowpea showing high leaf water content after 48 hrs. (Walker and Miller, 1986).
8. 1-2 week earlyness with high yield in cowpea under high temperature and drought conditions.
9. Higher values of WUE, LA and xylen ABA with lower DELTA (C_{13}) values in cowpea can be used for screening under drought situations (Nagugi *et al.*, 1994).
10. Larger tap root system under wilting conditions in India, was an indication of drought tolerance in cowpea (Itani, 1992).
11. A shoot: root ratio (0.30-0.50) in tolerant than is susceptible (0.70-0.90) cowpea lines during extreme dry situations of less than 1% soil moisture, was found most pertinent parameter for drought tolerance in Indian arid situation.

(iv) Breeding

A plant breeder, whose prime responsibility is to develop drought resistant genotypes, must take and follow the multidisciplinary approach in evaluation of large germplasm as the first stept towards drought tolerance breeding.

Collaboration with crop Physiologists is the most crucial as he brings to the breeding efforts a knowledge of plant mechanism and experimental techniques that are necessary for evaluation, the same may however, be modified and repatronized time to time for easy, early and lasting screening of breeding lines. Some papers old and recent define the methods of breeding crops for drought resistance but, none specifically defines methods for cowpea evaluation.

Generally plant survival has been used as a measure of drought resistance by crop Botanists but Breeders and Agronomists prefer high productivity as the measure of drought resistance, because survival does't many a times, contribute towards productivity. Often plant survival in moisture deficit situations is achieved by cryptobiosis: reductions in plant stature, leaf area, leaf dry weight, reduced metabolic activities which directly affect the plant productivity. Population adaptation for increased productivity is however, effectively judged when plants are grown under natural moisture deficit field conditions which is, however, subjected to environmental variability.

Drought resistance in crop plants is generally due to avoidance capacity. Most of these avoidance traits are xeromorphic. However, true drought resistance is not possible without the development of xeromorphic traits provided resistance.

Cowpea also falls in xeromorphic group of adaptation, characterized with possession of inherent avoidance type tolerance, particularly in terms of fast initial shoot and root growth, broader leaves, medium maturity and few pods with low productivity and indeterminate growth habit.

Cowpea a typical legume is adapted to arid as well as semi-arid situations from low (<250 mm) to high (>600 mm) rainfall situations on loamy sand to sandy loam soils. It therefore, encounters with two types of soil moisture situations stored moisture and variable moisture environment for completion of its growth. Thus, specific plant types for these situations are required as below:

Stored soil moisture: Rapid uptake of soil moisture, storing water in tissues and retarding water loss through epidermis and stomata. Increasing metabolic efficiency, rapid growth and deep root system.

Variable soil moisture environment: Plants having high photosynthetic rate, stomatal sensitivity, rapid osmotic adjustment, diverse root system, may not necessarily be deeper one and inderminate growth habit.

Cowpea is possessing adaptive traits for aforesaid both situations, hence, showing geographic distribution to both conditions. For instance, cowpea adjudged for high photosynthetic capability, is a known dehydration avoider with high stomatal sensitivity, rapid growth, deep as well as dense root system and rapid uptake, thus providing it avoidance type tolerance mechanism. It however, requires better source-sink relationship, early partitioning and remobilization of carbohydrates from dense canopy and broad leaves to reproductive parts.

Looking at the growing habitates and exemplary evidence of moro-physiological traits associated with cowpea and our experiences on other crops of drought habit, it may conceiveably be argued that selections for earlyness followed by improved avoidance may by far, prove the most successful breeding strategy for this crop. Early maturity representing escape mechanism would be helpful in bringing stability in yield and reducing the risk of crop failure in completion of life cycle during soil moisture pecipitation only. More important is that earlyness may be associated with yield enhancement also as evident from existing reports in cowpea and moth bean, the sister arid legume. Thus, breeding/selection for earlyness with improved yield potential could be useful in achieving the efficient water use in cowpea (Mohammed *et al.*, 2000).

Cowpea being an indeterminate crop is difficult to be improved in earlyness because early flowering and maturity both are strongly influenced due to varing environment. Early flowering will lead to improved source-sink relationship. Selections for earlyness may however, be attempted on targeted situations only and particularly, on wider spacing situations in which cowpea is commercially planted. More interesting and valuable improvement in yield would be realized by curtailing the vegetative phase and enlarging the reproductive phase within defined life cycle in a particular soil moisture zone. Due to continued poding, it may not be ideally suitable to select early maturity which may also be influenced due to sharp soil moisture deplection due to terminal drought, hence maximum number of mature pods at the early stage may prove a sound strategy.

Hybridization between traditional varieties representing avoidance mechanism of tolerance with early maturing ones carrying escape phenomenon, may give segregants, combining genes both for escape and avoidance mechanisms hence, showing adaptation to long range of environments.

Generally, if the adaptation to specific environment is concerned, plants adapted to optimum conditions may not show similar adaptive traits in drought situations. Thus, this philosophy provides opportunity for operation of selection in stress conditions only. Plant types may therefore, have distinguished types in two situations. However, is term of variable environment, mediocritic plant behavior may do well in selection of desired types. Thus, prime concern in breeding for drought tolerance is attempting selection or screening of germplasm which is most pertinent to specific situation and satisfying the desired type of tolerance.

While screening for drought tolerance, it is necessary that high productivity of a cultivar or residual high productivity of recently bred variety in dry region may not be treated as drought tolerance. High yield and resistance potential to drought are by and large considered as the independent traits. Thus, drought tolerance on targeted situation and high yield in optimum environment may be assessed in succession. Proportional reduction in economic traits or the drought related morpho-physiological traits in drought over control conditions may give some idea about the genetic variations related to drought existing in a given population. For effective screening, very high selection pressure so as to sieve out 90-95% susceptible lines would be quite appropriate. Selection of plants on soil moisture reduced to 1-2% availability, may prove appropriate in isolating ideal lines showing inherent drought potential. Furthermore, selection on targeted drylands during usual cropping season in wider row spacing (where crop plants are to be grown) at the reproductive stage may provide desired results.

The most appropriate strategy on improving drought tolerance potential of cowpea plants by and large depends on the collection of appropriate germplasm from real habitat on cultivators fields representing both hot and dry conditions. The same are tightly confounded with each other, one contributing to other and proving supplimentry to each giving compound and complicated response and are difficult to separate out. Selection of plants during noon or at the peak hour of temperature on to drylands when they are showing tendency of wilting and folding back of leaves, would result in drought tolerant plants (showing no leaf folding). Sometimes plants showing mediocritic behaviour may be preferred for this purpose.

(v) Drought Tolerant Varieties

Attempts on improving variability for tolerance to drought in crops including cowpea have been put up since past century or even before, yet true drought tolerant cultivars being exploited commercially have not come to the fore. This indicates complexity in the systems in its expression, utilization and ultimately in transfer. Reported tolerant genotypes listed in table 1 indicate the results of collection of germplasm in optimum or sub-optimum situations. A yielding genotypes (s) was referred to as drought tolerant which in true sense of resistance is not. To be precise no variety, deliberately proposed to be developed as drought tolerant was bred as having true physiologically drought resistance (without escape and/or avoidance). Some of the germplasm lines found tolerant to drought at this Insitute are depicted in Plate 1.

Table 1. Drought tolerant genotypes of cowpea

S.N.	Genotype	Country	Reference
1.	Katumani-80 (dual purpose)	Kenya	Shakoor *et al.*, 1984
2.	New era	Nigeria	Babalola, 1980
3.	Rogue	Norway	Aune, 1992
4.	KVX 30-305-36	Norway	Aune, 1992
5.	Dana 11a	Nigeria	Singh *et al.*, 1999
6.	TVU – 11979	Nigeria	Singh *et al.*, 1999
7.	IT 90K–59-2	Nigeria	Singh *et al.*, 1999
8.	EPACE-1	India	El Maarouf *et al.*, 1999

9.	CoUV-702	India	Anonynous, 2000
10.	IGFRI - 1440	India	Yadav and Patil, 1984
11.	HFG-42-1	India	Yadav and Patil, 1984
12.	Severe II	Nairobi	Muruli *et al.*, 1980
13.	Mak 1/5/B	Nairobi	Muruli *et al.*, 1980
14.	EG-3	Philippines	—
15.	TVa-4-b-lab	Philippines	—
16.	IC-39856	India	Kumar and Seema Gaur, 2002
17.	IC-39853	India	Kumar and Seema Gaur, 2002
18.	IC-39890	India	Kumar and Seema Gaur, 2002
19.	Co-4	India	Selvaraj *et al.*, 1986.

SALT

Salinity is basically the problem of irrigated agriculture, showing its build-up in the soil and affecting plants multi-dimensionally. The problem of salinity is more severe in arid and semi-arid regions where salts accumulated naturally or through irrigation water, show upward movement through capillary rise, are redistributed on the upper surface of the soil. The rainfall or the good quality water are not adequate to leach out the salts from the effective root zone. Principally, the problem precipitates due to continued waterlogging due to lack of provision of effective drainatge of standing water. The same is the secondary cause of salinitization in canal command areas or in areas characterized with high water table. Salinity resulting from the water-soluble salts in different proportion may lead to slow but steady desertification of the soil. Saline soils are virtually treated as unfit for commercial agriculture as the salts may show varied impacts on emergence, growth, development, physiological functions, total produce and the quality of produce as well. Salts may affect plants through osmotic effects, ionic effects and ultimately due to the toxicity of ions. Salinity may also produce effects on plants through water scarcity. The problem of salinity can be considered more serious to drought because the latter can be managed through providing irrigation in drought prone areas, besides being costly affair, may additionally lead to salinization in future course. Moreover, the salt content may show extreme variation in root zone which makes it difficult to find out a uniform patch of the soil for effective screening against this menace. That is why, screening for salt tolerance is more difficult than for drought. More serious problem with salinity is its varied impacts on different growth stages; the adverse effects of salinity are however visualized greatly at seed germination and seedling emergence stages leading to sub-optimal plant stand. Dressing the crops with inorganic fertilizers (particularly, nitrogenous fertilizer) may not prove helpful in improving the plant growth. Thus, management of salt problem is much more difficult, skill based and requires adequate and specific resources. Some-times problem of salinity is confounded with alkalinity and saline-alkaline situations, which may require altogether different management strategies.

Salinity is a natural consequent, men made mismanagement, hence its mechanical management is very difficult, cumbursome and costly. Biological approache: use of halophytes, by growing salt tolerant genotypes of concerned crops within their tolerance limit, is by far the most convenient and economic approach being socially adopted by masses.

Cowpea an important legume grown for food, fodder, green manuring, is cultivated in arid and semi-arid regions as well. In northern western parts cowpea is grown as kharif rainfed but in Southern India it is grown in winter season also. Thus, due to wide range in its geographic, seasonal and climatic, conditions for cultivation, there is every likelyhood that cowpea will be exposed to salinity hazards also. It has potential to be grown in all three seasons of tropical Indian sub-continent. Cowpea being a pulse is likely to be adversely affected with the salinity problems, hence, remedial measures for its cultivation with respect to salinity problem are strongly desired.

Salt tolerance

From the extensive researches directed during the past half century, world over, following inferences can be drawn on plant tolerance to salts:

1. Problems of salt adversity in reference to soil appears to be quite variable over the space and time and perhaps more complex than other abiotic stresses in crop x soil x climate compendium.

2. Tremendous variations at inter and intra-crop level exist in respect to salt tolerance.

Salt tolerance is the ability of plants to grow and yield satisfactorily. However, Levitt (1980) has used salt tolerance in a broader perspective.

1. Salt tolerance: When plants respond to salinity by accumulating salts in their cells or in specific cellular parts-salt hairs or salt glands etc.
2. Salt avoidance: When plants avoid salt stress by keeping the internal concentration unchanged by water absorption or by avoiding salt uptake.

Following the classification of crops in different classes of tolerance, using thereshold salinity level, cowpea with ECe 4.9 $dS.m^{-1}$ as a threshold limit, is rated moderately tolerant (Maas and Hoffman, 1977). Among pulses soybean also falls in this class.

Cowpea however, proved the most sensitive to salinity at the vegetative stage and the tolerance increases with increasing plant growth. Seed yield reduced by 50% at the soil solution having OP of –0.45, -0.76 and -0.88 MPa in cowpea plants salinized during vegetative, flowering and podding stages, respectively, at Riverside situations (Maas and Poss, 1989). Thus, cowpea can tolerate salts at the reproductive stage to which majority of the crops are said to be sensitive.

N-fertilized cowpea like other legumes may prove more tolerant to salt stress compared to those dependent on N-fixation only. This was attributed that fertilizer N alleviates the inhibitory effects of NaCl and Na_2SO_4 on N_2ase activity. Nodulation including nodule number, their dry weight and nitrogenase activity are decreased in cowpea in response to salinity. For instance, in a study at California, with cowpea cv. Black eye No. 5, root tended to be more fibrous and nodules were smaller in size and number above 7.5 pH (Joe and Allen, 1980). Similarly, process of nodulation was very salt sensitive in gram than in soybean.

Mechanism

Salt tolerance mechanism of pulses, particularly, cowpea may not be different than other field crops which have been extensively investigated. The common mechanisms have been selectivity of ion transport and their accumulation in specific tissues which may contribute as a gradiant for osmotic adjustment and avoidance of salt absorption or selective absorption of less injurious ions (K, Ca) than deliterious ones (Na^+, Cl^- etc.). However, there is no published report on mechanism of salt tolerance operating in cowpea, but one or two reports on *Vigna* group are available which are summarized below.

In moong bean cv. Barker grown with sand culture at 0, 50, 100, 150, 200 and 250 mM NaCl, very effective exclusion of Na^+ and Cl^- ions was observed (Saleem and Pitman, 1987). Conribution of proline as osmotic gradiant was however, of minor importance in legumes inclucing *Vigna* group of species (Larher *et al.*, 1987).

SDS-page polypeptide profiles of a salt sensitive and salt resistant leaf derived callus grown in the presence or absence of 300 mol^{-3} NaCl, revealed that three polypeptides of 35, 46 and 80 KDa were salt induced. The former two were detected in salt stress callus of both lines, while latter was unique with the salt resistant line in moong bean. Synthesis of 5 other polypeptides of 158, 138, 131, 69 and 42 KDa decreased in salt resistant lines during salt stress. Thus, proteins induced/ reduced during NaCl treatment may play a part in salt resistance as observed in moong bean (Gulati and Jaiwal, 1993).

Tolerance of tropical legumes

The tolerance of a range of tropical grain legumes to salinity was compared during early vegetative growth of plants grown in pots with NaCl salinity having ECe value of 1.3 to 13.8 dSm^{-1}. Tolerance based on the ECe for 50% of maximum growth, was of the order: *Sesbania cannobins* (13.2 dSm^{-1}), guar cv. CP 177 (10.0 dSm^{-1}), guar cv. Brooks (9.8 dSm^{-1}), cowpea cv. Caloona (9.0 dSm^{-1}), soybean cv. CPT-126671 (6.7 dSm^{-1}), pigeon pea cv. Hunt (5.4 dSm^{-1}), black gram cv. Regur (4.9 dSm^{-1}), green gram cv. Clera (3.5 dSm^{-1}) (Keating and Fisher, 1985). Cowpea maintainaned low Na^+ in shoots but accumulated moderate levels of Cl^- in presence of Cl^- salts. There was no evidence that ion excess was a limiting factor of growth in cowpea and gram pulses (Keating, 1986).

Seedling growth of cowpea was reduced more with salinity (EC 2.5-10.0 dSm^{-1}) than alkalinity (SAR 10-20). Cowpea cv. FS-68 showed the higheat tolerance (Singh, 1980).

Growth was inhibited in cowpea (cv. Early Season) when 25 mM NaCl was added to medium at imbibition but 10 mM NaCl promoted growth. Process of nodulation was more salt sensitive in cowpea than in moong bean and soybean crops.

The increasing levels of salinity (3, 6, 9, 12 and 18 dSm^{-1}) by applying NaCl and $CaCl_2$ (in 4:1 ratio) decreased germination and seedling growth. Most of the cultivars of cowpea and pigeon pea tolerated salinity in range of 6-12 dSm^{-1}, cowpea was more sensitive to pigeon pea in Udaipur situations (Paliwal and Maliwal, 1973).

In cowpea, the vegetative yield was more decreased than dry seed yield by increasing soil salinity. For instance, vegetative growth was reduced by 9% for each unit increase in salinity of the soil extract beyond a therehold value of 1.6 dS m^{-1}. Dry seed yield was however, reduced by 12% beyond 4.9 dSm^{-1} (West and Francois, 1982). Germination was significantly reduced in sand culture exceeding 12.0 dS m^{-1} at River-side In Indian situations, germination of cow was significantly reduced at 6.23 dS m^{-1} salinity. Cowpea was, therefore, sensitive at germination but tolerant at vegetative growth stage.

However, in terms of soil sodicity, cowpea proved sensitive, wherein cowpea germinations was reduced at ESP 44.2 whereas, fresh yield and leaf succulence decreased at 30 and 60 days of sowing, respectively (Singh *et al.*, 1980).

The length, fresh and dry weight of the enbryo aris of cowpea decreased with increasing salinity levels.

The adverse effects of alkalinity was associated with a decrease in chlrophyll, protein, DNA and RNA contents in cowpea which led to the sensitivity of this crop to alkalinity (Gill, 1979). Moreover, adverse effects of cowpea root zone alkalinity were on branches, stems and leaves (Gill, 1987). At higher range of sodicity, there was marked red-uction in dry matter yield, number of nodules and nitrogen content of cowpea but substantial growth and nodulation occur-red upto 34 ESP, further number of nodules declined more sharply than dry matter yield (Batra and Bhardwaj, 1981).

Breeding for Salt Tolerance

Strategies on breeding for salt tolerance in cowpea may be adopted as far normal situations. The real difference is of effective and viable screening methodology, build-up of adequate salt content in the soil and identification of tolerance based morpho-physiological traits by use of adequate selection pressure. In view of complexity of problems and the experiences gained in other crops, following probable breeding steps can be suggested:

1. *Germplasm collection from the salt habitats:* Cowpea grown on moderately saline soils, with poor quality supplimentry irrigation or in extreme water scarcity situations may be used for this purpose. Cowpea germplasm being grown with absolute saline situations may not be available, but osmotic stress being common to both (drought and saline) situations, lines selected for extreme drought habitats may behave well in saline situation also. Thus, germplasm collection is most important strategy, as practical results have been generated in many crops by using this approach. The lines being grown on salt habitats for long have gradually adapted by synthesis of need based salt shock proteins leading to their required adaptation for survival value, without being improved in yield potential.
2. *Screening:* Collected germplasm, breeding lines or related material may be screened in phased manner in small micro-plots with synthetic saline waters of desired quality. The water may at least be of three concentrations, representing one threrhold level and one lower and one higher to it. The quality and concentration of saline water may represent composition of underground water of the surrounding local areas. The screening may not at all be done on saline lands, the same being highly heterogeneous hence, may not represent uniform saline patch. Cowpea is understood sensitive to vegetative stage, hence, evaluation at seed germination and at seedling stage may be done precisely. Levels of salinity may be used as 0,6,9 and 12 dSm^{-1}; as ECe 9.0 dSm^{-1} is

regarded as the threshold salinity level for cowpea. Comparison of salt tolerance may be done on the basis of per cent increase/decrease value of the morphological traits in saline over control treatment. Lines representing absolute high grain yield in saline conditions only and others with proportionally lesser proportion of reduction over control may be isolated. Lines improved in germiation potential *per se* may also be identified.

Plants showing wider K : Na ratio at the reproductive phase, less pollen sterility at higher salinity level, comparatively faster growth of plants, less tip burning may indicate some signs of salt tolerance.

Salt tolerant varieties reported by different quarters from within existing germplasm lines or the commercial cultivars, showing better performance are given in Table 2.

Table 2 Salt tolerant genotypes of cowpea

Genotype	Tolerance limit	Country	Reference
Vita	0.1% M NaCl	Brazil	Melo *et al.*, 1994
C-152	18-22 ESP	India	Gill, 1983
Caloona	9.0 dS. m^{-1}		Keating and Fisher, 1985
FS-68	SAR 10-20	India	Singh, 1980

HEAT

Heat stress due to ambient temperature higher than optimum, is regarded as the chief cause, decreasing crop productivity in arid areas; the yields realized in such situations are far less than the potential average yield of the concerned crop. High temprature (>30°c) generally limits the growth and adaptation of legumes in many countries (Brown, 1960, Keteing, 1984, Werry *et al.*, 1994), proving, therefore, a major cause of low and unstable yield. Adversity of heat stress is likely to be accentuated in future cource due to continued global warming (Schneider, 1989). Improvement in heat tolerance in legumes may contribute to production sustainability and provide a means of extending legumes cultivation to previously unsuitable regions and in the untraditional seasons.

In arid regions, soil and air invariably remain dry, therefore, lack of moisture and humidity in the air disfavour the growth of plants being grown with limited resources. In such situations, heat becomes an important abiotic stress, where the plants experience uncomfert for usual growth.

Productivity of cowpea is considerably reduced due to heat (Hall and Patel, 1987, Nielson and Hall, 1985b Warang and Hall, 1992, 1993) either by enhancing the adverse effects of limited water or independently. The combination of high night temperature and longer days can considerably inhibit or slow down the floral bud development (Dow-El-Madina and Hall, 1986; Patel and Hall, 1990) in legumes. Also Nielson and Hall (1985b) demonstrated that minimum night temperature (>20°C), being common in majority of cowpea growing areas (Nielson and Hall, 1985a) can result in poor pod set due to increased pollen sterility. Low pod set under heat prone zones may result from sterility associated with anther indehiscence and incomplete pollen development. Precisely the productivity of cowpea can be reduced considerably due to hot weather prevailing during flowering stage (Margo and Hall, 1992).

Cowpeas being grown during rainy season in Northern-Western India and during winter season in Southern India are encountered with rising temperature and fast depleting soil moisture towards maturity. Thus, terminal drought, accompanied by sub-optimal high temperature can cause serious reduction in cowpea productivity through various physio-bichemical limitations. Possibility of providing irrigation and lowering the atmospheric temperature during peak of cowpea crop are remote. Hence, remedial measures, particularly biological means through identification of heat tolerant genotypes is the key issue in realization of fullest potential of cowpea in arid and semi-arid regions of India.

Response

Morphological: Plant response to high temperature are diverse including protein denaturation, change in lipid composition, reduction in membrane stability, decreasing photosynthesis, cessation of cytoplasmic streaming and reduced yield through high extent of sterility. These changes being diversified include morphological, biochemical and physiological traits, varying greately with day/night temperatures and their duration.

Generally high temperature stress causes accelerated plant development resulting in reduction of growth and grain yield (Blum, 1988). When pot grown cowpea experienced large diurnal variation in root temperature (33-19°C) during reproductive growth, floral morphlology was markedly affected, petals were constructed in such a way as to restrict self pollination, hence seed yield was drastically reduced. High temperature may cause reduction in germination of cowpea seeds. Findings of such study indicated that 12 to 40°C temperature gave a harmful impact on two cowpea cultivars, but 32°C temperature was observed as optimum for better germination of cowpea seeds. Nodule activity was adversely affected due to change in number of peduncles/plant as the temperature increased from 25°C to 40°C (Minchin and Summerfield, 1976).

Stage: Heat may produce more drastic effects on the young plants, as the seedlings may be more severely affected than the matured plant. Productivity of cowpea could be reduced substantially by hot weather during flowering stage (Margo and Hall, 1992). High temperature injury varied considerably amongst the growth stages, being highest at the flowering stage whereas, susceptibility to injury decreased towards maturity. Generally high temperature (35/30°C) affected flowral morphology by increasing hypathium length causing stigma exertion, reducing the chances of fertilization.

Biochemical: At the cellular level, high temperature can lead to protein and membrane degradation, depletion of carbohydrate reserves due to destabilization of photosynthesis II functioning, consequently inhibiting the process of photosynthesis. Heat may also cause inactivation of enzymes, damage of cellular substances. At maturity, heat inhibits the accumulation of starch into grain, therefore, reducing crop productivity (Jenner *et al.*, 1991). Setting of pods at high night temperature may be more at shorter than longer photoperiods with a critical photoperiod of 12-13 hrs (Mutters, 1989). Heat does not affect the plant as a whole, but individual plant cells receive a strain upon it, because heat prohibits certain enzymes to work, delocalization of cellular components and denaturation of protein which leads to the death of cells. High temperature induced yield losses could also be due to accelerated senescence, reduction in photosynthetic rate and increased respiration.

Photoperiod: Not only the day temperature but the night temperature and the length of day and night as well are equally important in influencing the adaptation and productivity of plants. Adaptation to day length is equally important in cowpea but night temperature also affects response to day length and may even be more important in the determining flowering time and seed yield (Huxely *et al.* 1974). Warm day temperature accelerates the rate of leaf senescence in cowpea and also considerably promoting abortion of peduncle, flowers and young pods.

Cowpeas are rarely insensitive to photoperiod but are typically short-day plants, wherein, photoperiods longer than the critical i.e. 12.5 to 13.5 h d^{-1} (Dow-el-Madina and Hall 1986, Craufurd *et al.*, 1996) increase the number of days to flowering (Hadley *et al.*, 1983, Craufurd *et al.*, 1996). Both photoperiod-sensitive and insensitive cowpea genotypes flower sooner in warmer than cooler season.

The grain legume cowpea exhibits considerable genetic variations in response of flowering to photoperiod regimes (Hadley *et al.*, 1983, Craufurd *et al.*, 1996). The implication of these responses to photoperiod and temperature for adaptation of genotypes to varied sowing dates are considerable.

Tolerance: For high temperature tolerance, heat tolerant cv. need to have more heat stable photosynthetic system, greater thermostability in heat sensitive enzymes, and membranes. Tolerance at early stage is crucial for early screening. However, in cowpea, lines are available showing heat tolerance during flowering but are susceptible during pod set (Ehlers and Hall, 1996). The night temperature in range of 25-30°C may show substantial damage to subsceptible lines during floral bud development and anther developments stages (Mutters, 1989, Ahmed *et al.*, 1993). However, interesting findings revealed that heat tolerant genes imparting resistance at early flowering and pod set may also contribute equally to final grain yield through their uniform effects on pod set and harvest index (Abdelbagi and Hall, 1999).

Indices: Morphological features such as light green foliage, vertically oriented, pubescent and waxy leaves that reduce heat absorption, heat shock proteins for thermo-protection and high total non-structural carbohydrates coupled with an efficient partitioning system for root and shoot growth and viability are required.

At maturation of pollen grains, heat sensitive genotypes had lower proline concentration in the pollen grains and was higher in anther walls than heat tolerant genotypes under hot situations (Mutters *et al.* 1989).

Mechanism: High temperature tolerance in plants is achieved when tissue temperature is high and the plants are still surviving. Like drought, plants tolerate heat by way of avoidance and tolerance mechanisms. Avoidance mechanism keeps the canopy temperature lower than the ambient temperature through many ways: transpirational cooling, reflection of solar radiation through wax deposition or hairy surface. The tolerance against high temperature is achieved by three ways: biomembrane saturation, its protection and sysnthesis of certain enzymes like peroxidase, superoxide dismutase. In such situations, HSPs may play significant role. Heat shock proteins have not however, been reported in cowpea or efforts have not been directed in this direction.

Screening techniques: A rapid yet effective screening technique for isolating heat tolerant plants at the early vegetative stage in any reasonable environment is required that would allow screening of multiple generations and large number of germplasm in a short period, leading to acceleration in breeding progress towards heat tolerance (Abdelbagi and Hall, 1999).

A number of screening indices in different crops including cowpea are being employed for isolating heat tolerant genotypes. For instance, high amylase activity at the seedling stage, accumulation of HSPs, pollen germination, massive proline translocation towards pollen grains (Talwar *et al.*, 1999), superoxide dismutase activity, heat killing temperature, electrolyte leakage rate etc. have been used. Moreover, longer and thicker leaves, thicker paliside tissue, more leaves/plant, taller and narrow plants and higher chlorophyll a/b ratio have been used as the covenient morphological traits. The response to heat are varied hence, precisely following methods can be used in context of cowpea.

1. Leaf electrolyte leakage as a measure of thermo-stability may provide an efficient indirect screening

technique for reproductive-stage-heat tolerance genes that can be used for the plants grown in a wide range of nursery environments.

2. High night temperature during late floral stage can impair pod set by causing anther dehiscence and reduced pollen fertility. Screening of cowpea for flower production and pod set in extremely hot field conditions with high night temperature (Hall, 1992, 1993) may, therefore, prove quite useful.
3. Assessment of the induction temperature and high stringency lethal stress levels, which involve the optimization of stress induction temperature, when stress responsive genes are expressed (Kumar *et al.*, 1999).
4. The most convenient approach for screening against heat tolerance is counting the number of mature pods or the number of grains per pod towards maturity under high day temperature.

Breeding: Limited progress achieved so far in breeding for heat tolerance in legumes including cowpea is due to imperical approach. Inspite of greate surge in morphological and physiological approaches their application in heat tolerance appears quite modest. Further more, a negative feature in this regard is the low heritability for end product of grain yield in hot and dry climate (Blum, 1988). Thus, development of screening techniques which are trait based are strongly desired.

Major genes for heat tolerance were observed to be involved in *P. vulgaris* (Bouwkamp and Summers, 1982). Similiarly Dickson and Petzoldt (1989) also observed that heat tolerance was a dominant character with low heritability (0.14) in *P. vulgaris*. On the basis of pods/ peduncle and the proportion of tolerant plants in F_1, F_2 and back cross progenies, heat tolerance in cowpea was conferred by a single dominant gene (Table 3). In view of it, pedigree method of breeding with selection in F_2 is suggested. Breeding strategies for heat tolerance on the basis of existing information and the available plant material can be set as follows:

Table 3. Segregation of F_2's for pods per peduncle from crosses among heat-tolerant and heat-sensitive cowpea

Population	Observed segregation		% Tolerant progeny	Expected ratio sensitive: tolerant	X^2	P-value
	Sensitive	Tolerant				
Tolerant x sensitive			1987			
Prima x Magnolia	17	65	79	1:3	0.79	0.25-0.50
Prima x Bambey 23	25	74	75	1:3	0.00	>0.09
Prima x 7964	29	64	69	1:3	1.89	0.10-0.25
Homogeneity					0.12	0.50-0.75
Tvu 4552 x Magnolia	40	90	69	1:3	2.31	0.10-0.25
Tvu 4552 x Bambey 23	41	60	59	1:3	13.09	<0.01
Tvu 4552 x 7964	22	44	67	1:3	2.44	0.10-0.25
Homogeneity					3.00	0.10-0.25
Homogeneity (Prima Crosses with Tvu 4552 Crosses)					11.40	0.10-0.25
Tolerant x tolerant						
Prima x Tvu 4552	15	64	81	All tolerant		
Sensitive x sensitive						
7964 x Bambey 23	102	31	23	All sensitive		
7964 x Magnolia	79	56	42	All sensitive		
Bambey 23 x Magnolia	51	3	6	All sensitive		
			1988			
Tolerant x sensitive						
Prima x sensitives	66	154	70	1:3	2.93	0.05-0.10
Tvu 4552 x sensitives	36	83	70	1:3	1.75	0.10-0.25
Homogeneity					0.00	>0.99
Tolerant x tolerant						
Prima x Tvu 4552	9	66	88	All tolerant		

1. For most crops, only a small segment of germplasm has been collected and evaluated basically for heat tolerance only. Future course of improvement is highly

dependable on the flagging a trait for heat tolerance and using this trait for collection of germplasm from the natural habitat representing high temperature at the critical growth stage (pod formation) and its evaluation for heat tolerance and yield assessment in targeted environment.

2. In view of inability to control temperature in open fields and to separate out the effects due to heat and drought, evaluation and selection of plants become all the more very difficult demanding, therefore, very strong morpho-physiological simple techniques. The techniques which may define plant avoidance and tolerance separately are required.

The most effective and reliable method is electrolyte leakeage test in controlled conditions during early stage indicating stability of the membrane, reflected through plant vigour and morphology. Heat affected plant productivity through increased pollen sterility, hence, examination of the same may throw some promising lines.

Heat tolerance is a dominant character with low heritability, thus, transfer of heat tolerance genes in an otherwise agronomically superior genotype through pedigree method in F_2 generation may not pose serious problem. Heat tolerance source in cowpea are already available – 779, Prima, TVU-4552 and IC-39888.

REFERENCE

Abdelbagi, M.I. and Hall, A. E. 1999. Productive-stage heat tolerance, leaf membrane thermostability and plant morphology in cowpea. *Crop Sci.* 39: 1762-68.

Aggarwal, V.D. 1987. Selecting cowpea varieties for harsh, semi-arid sahelian areas. *Tropical Agric.* 8(2): 2.

Ahmed, F.E., Mutters, R.G. and Hall, A.E. 1993. Interactive effects of high temperature and high light quality of floral bud development in cowpea. *Australian J..Pl. Physiol.* 20(6): 661-67.

Anonymous 2000. Annual Report, Central Arid Zone Research Institute, Jodhpur.

Aune, J. B. 1992. Cowpea cultivars for rainfed cultivation in Gourma, Mali. *Nigerian J. Agric. Sci.* 6(2): 145-160.

Babalola, O. 1980. Water relations of three cowpea cultivars. *Pl. & Soil* 56: 59-69.

Batra, Lalita and Bhardwaj, K.K.R. 1981. Growth, nodulation and nitrogen fixation in legumes as affected by soil sodium. *J. Indian Soc. Soil Sci.*29: 530-536.

Biswas, A.K. and Choudhuri, M.A. 1986. Dikegulac sodium-a potential water stress hardening agent. *Indian J. Exp. Biol.* 24: 608-11.

Blum, A. 1988. Plant breeding for stress environments. CRC Press Inc. Boca Raton, Florida.

Bouwkamp, J.C. and Summers, W.L. 1982. Inheritance to resistance to temperature-drought stress in snap bean. *J. Hered.* 73(5): 385-86.

Boyer, J.S. 1996. Advances in drought tolerance in plants. *Adv. Agron.* 56: 187-218.

Brown, D.M. 1960. Soyabean ecology. 1.Development temperature relationship from controlled environment studies. *Agron. J.* 52: 493-96.

Chaturvedi, G.S., Aggarwal, P.K. and Sinha, S.K. 1980. Growth and yield of determinate and indeterminate cowpea in dryland agriculture. *J. Agric. Sci.* (Camb.) 94: 137-144.

Craufurd, P.Q., Roberts, E.H., Ellis, R.H. and Summerfield, R.J. 1996. A stability analysis of time to flowering as a screening for responsiveness to temperature and photoperiod in cowpea *(Vigna ungiculata* L.) *Euph.* 88: 77-84.

Dickson, M.H. and Petzoldt, R. 1989. Heat tolerance and pod set in green beans. *J. Am.. Soci. Hort. Sci.* 114: 833-836.

Dow-El-Madina, I.M. and Hall, A.E. 1986. Flowering of contrasting cowpea *(Vigna unguiculata* (L.) Walp) genotypes under different temperatures and photoperiods. *Field Crops Res.* 14: 87-104.

Dudal, R. 1976. Inventory of the major soils of the world with special reference to mineral stress. In: *Plants Adaptations to Mineral Stresses in Problem Soils* (Eds. M.J. Wright), Cornell Univ., Agril. Exp. Sta., Ithaca NY-3.

Eckholm, E.P. 1975. Salting the earth. *Env.* 17:9.

Ehlers, J.D. and Hall, A.E. 1996. Genotyphic classification of cowpea based on responses to heat and photoperiod. *Crop Sci.* 36(3): 673-79.

El-Maarouf, H., Juily-Fodil, Y., Gareoil, M.D., Arcy-lameta, A. and Arh-Thu-Phan-Thi 1999. Enzymatic activity and gene expression under water stress of phospholipars D in two cultivars of *V.*

unguiculata L. Walp, differing in drought tolerance. *Pl. Molecular Biol.* 39(6): 1257-65.

Franca, M.G.C., Anh-Thu-Phan-Thi, Pimentel C., Rossiello, R.O.P., Zuilytodil, Y. and Laffray, D. 2000. Differences in growth and water relations among *P. vulgaris* cultivars in response to induced drought stress. *Env. and Exp. Bot.* 43(3): 227-37.

Gill, K. S. 1983. Note on alkali tolerance of cowpea. *Curr. Agric.* 7(1-2): 82-84.

Gill, K. S. 1990. Effect of saline irrigation at various growth stages on growth, yield attributes and ionic accumulation pattern in green gram *(P. radiata). Indian J. Agric. Sci.* 60: 280-84.

Gill, K.S. 1979. Chlorophyll, protein, DNA and RNA contents and yield of cowpea and *dhaincha* grown at different levels of alkalinity. *Indian J. Agric. Sci.* 49: 7-12.

Gill, K.S. 1987. Effect of soil alkalinity on growth, yield and some biochemical parameters at grain filling stage in cowpea and *Sesbania cannebina. Indian J. Pl. Physiol* 30: 38-43.

Grantz, D. A. and Hall, A.E. 1982. Earlyness of an indeterminate crop *V. unguiculata* (L.) Walp as affected by drought temperature and plant density. *Australian J. Agric. Res.* 33: 531-40.

Gulati, A. and Jaiwal, P,K. 1993. Salt induced polypeptides in two callus lines of *V. radiata* (L.) Wilczek which differ in salt resistance. *Nat-Acad. Sci.* 16(11-12): 287-92.

Hadley, P., Roberts, E.H. and Summerfield, R.J. 1983. A quantitative model of reproductive development in cowpea *(V. unguivulata* (L.) Walp) in relation to photoperiod and temperature and implications for screening germplasm. *Ann. Bot.* 51: 531-43.

Hall, A.E. 1992. Breeding for heat tolerance. *Pl. Breed. Rev.* 10: 129-68.

Hall, A.E. 1993. Is dehydration tolerance relevant to genotypic differences in leaf sequence and crop adaptation to dry environment? pp 1-10. In: *Pl. Resource* to *Cellular Dehydration During Env. Stresses*. (Ed. T.J. Close and EA. Bray). The Amer. Soc. of Pl. Pathologists. Maryland.

Hall, A.E. and Patel, P.N. 1987. Cowpea improvement in semi-arid regions of sub-Saharan Africa, pp 279-90. In. *Food Grain Prodc.* in *Semi-arid Africa.*(Ed. J.M. Menyonga). *Proc. Int. Drought Symp.* "Food Grain Production in Semi-arid Africa, Nairobi, Kenya, May 1986, SAFGRAD, Nairobi.

Huxley, P.A., Summerfield, R.J. and Hughes, A.P. 1974. The effect of photoperiod on development of soybean and cowpea cultivars grown is UK in summer. *Exp. Agric.* 10: 225-39.

Itani, J. 1992. Studies on water absorption ability of cowpea. *Japanse J. Tropical Agric.* 36(1): 37-44.

Itani, J., Utsonomiya, N., Shigenaga, S. 1992. Drought tolerance of cowpea 1. Studies on water absorption ability of cowpea (*V. unguiculata* (L.) Walp var. unguiculata) *Jap. J. Trop. Agric.* 36: 37-44.

Jenner, C.F., Ugalde, T.D. and Aspinall, D. 1991. *Australian J. Pl. Physiol.* 18: 21-26.

Joe, W.H.E. and Allen, J.R. 1980. Effect of soil pH on plant growth and nodulation of cowpea. *Commun. Soil Sci. & Pl. Anal.*11: 1077-85.

Jordan, W.R., Monk, R.L., Miller, F.R., Rosenow D.T., Clark, L.E., and Shouse, P.J. 1983. Environmental physiology of sorghum. 1. Environmental and genetic control of epicuticular wax load. *Crop Sci.* 23: 552-58.

Khalil, S. and Mandurah, H.M. 1989. Effect of water deficiency and kinetin on growth and nitrogen metabolism of cowpea plants. *J. Agron. & Crop Sci.*164: 93-99.

Keating, B.A. 1986. Influence of salinity of ionic concentration and yield of three tropical grain legumes. *Australian J. Agric. Res.* 37: 167-177.

Keating, B.A. and Fisher, 1985. Comparative tolerance of tropical grain legumes to salinity. *Australian J. Agric. Res.* 36: 373-84.

Ketring, D.L. 1984. Temperature effects on vegetative and reproductive development of peanuts. *Crop Sci.* 24: 877-82.

Kumar, D. and Seema Gaur, 2002. Cowpea germplasm with multi-stress tolerance potential identified for arid situations. 2nd *Intern. Cong. Pl. Physiol. Sustainable Pl. Productivity Under Changing Env.,* 8-12 Jan. 2003, IARI, New Delhi (abstract) S-4, P-35, P-135.

Kumar, G., Krishna prasad, B.T., Savitha, M., Gopalakrishna, R., Mukhopadhyay, K., Ramamohan, G. and Kumar, U. 1999. Enhanced expression of heat shock proteins in thermo-tolerant lines of sunflower and their progenies selected on the basis of temperature induction response. *Theor. Appl. Genet.* 99: 359-67.

Labanauskas, C. K., Peter, S., Stolzy, L. H. and Handly 1981. Proteins and free amino acids in field grown cowpea seeds as affected by water stress at various growth stages. *Pl. and Soil* 63: 355-368.

Larher, F., Huo, S.M.J. and Gerantsauvage, D. 1987. Salt sensitivity in legumes during at the early stage of development. *Colloques de I'I NRA*, 3: 181-92.

Lawn, R. J. 1983. Responses of four grain legumes to water stress in South western Quensland. IV. Interaction with sowing arrangement. *Australian J. Agric. Res.* 34: 661-69.

Levitt, J. 1980. Response of plants to environmental stresses. Vol. 2, 2nd Ed., A. P., N.Y.

Maas, E.V. and Poss, J. A. 1989. Salt sensitivity of cowpea at various growth stages. *Irrig. Sci.* 10: 313-20.

Maas, E.V. and Hoffman, G. J. 1977. Crop salt tolerance-current assessment. J. Irrig. *Drainage*, Div. Agric. Soc. Civ. Eng. 103: 115.

Margo, K.O. and Hall, A. E., 1992. Inheritance of heat tolerance during pod set in cowpea. *Crop Sci.*32: 912-18.

MC Cree, K.J. and Richardson, S. G. 1987. Stomatal closure vs. osmotic adjustment: a comparison of stress responses. *Crop Sci* 27: 539-43.

Melo, D. F., De Jolivet, Y., Facanho, A. R., Gomes Filho, E., Lina, M.S. and Dizengremoli, P. 1994. Effect of salt stress on mitrochondrial energy metabolism of *Vigna unguculata* cultivars differing in NaCl tolerance. *Pl. Physiol. Biochem.* (Paris) 32(3): 40: 5-12.

Minchin, F. R. and Summerfield, R.J. 1976. Symbiotic nitrogen fixation and vegetative growth of cowpea (*V. unguiculata* L. Walp) in water- logged conditions. *Pl. and Soil* 45: 113-27.

Mohammed, M.F., Ferreira, M.I. and Jones, H.G. 2000. Useful tepeary and cowpea lines in rationalizing water use for pulse legume production in Southern Egypt. *Acta Horticultature* 537: 813-22.

Morgan, J. M., Rodriguez-maribona, B. and Knights, E. J. 1991. Adaptation to water deficit in chickpea breeding lines by osmoregulation: relationship to grain yield in the field. *Field Crops Res.* 27: 61-70.

Mukherjee, S. P. and Choudhri, M.A. 1983. Effect of some reducing agents on water stress induced oxidative and deteriorative process of *Vigna* seedlings. *Biol. Plantar.* 25: 401-07.

Murli, B.I., Pathak, R.S., Mukunya, D.M., Karel, A.K., Keya, S.O. and Ssali, H. 1980. Cowpea research in Kenya. *Trop. Grain Legum Bull.* 19: 13-16.

Mutters, R. G. 1989. Reproductive physiology of cowpea (*V. unguiculata* L. Walp) at high temperature with different photoperiods. *Dissert. Abst. Intern. Sci & Eng.* 49: 8, 2955B.

Mutters, R.G., Ferreira, L.G.R. and Hall, A.E. 1989. Proline content of the anther and pollen of heat tolerant and heat-sensitive cowpea subjected to different temperature. *Crop Sci.* 29: 1497-1500.

Nagarajah, S. and Schulze, E.D. 1983. Response of cowpea to atmospheric and soil drought. *Australian J. Pl. Physiol* 10: 385-94.

Nagugi, E.C.K., Galwey, N.W. and Austin, R.B. 1994. Genotype x environment interaction in carbon isotope discrimanation and seed yield in cowpea. *Euph.* 73: 213-24.

Nielsen, C.L. and Hall, A.E. 1985a. Response of cowpea *(Vigna unguiculata* (L.) Walp) in the field to high night air temperature during flowering: 1. Thermal regimes of production regions and field experimental systems. *Field Crops Res.*10: 167-79.

Nielsen, C.L. and Hall, A.E. 1985 b. Response of cowpea *(Vigna unguiculata* (L.) Walp) in the field to high night air temperature during flowering. II. Plant responses. *Field crops Res.* 10: 181-96.

Obesesan, I.O. 1985. Association among grain yield components in cowpea. *V. unguiculata* (L.) Walp. *Genet. Agr.* 39: 377-80.

Paliwal, K.V. and Maliwal, G. L. 1973. Salt tolerance of some arhar and cowpea varieties at germination and seedling stages. *Ann. Arid Zone* 12: 135-44.

Patel, P.N. and Hall, A. E. 1990. Genotypic variation and classification of cowpea for reproductive responses to high temperature under long photoperiods. *Crop Sci.* 30: 614-21.

Robert, M., Zablatoueicz, D. Foeht, D. D. and cornell, G. H. 1981. Noduation and N-fixation of field grown California cowpea as influenced by well-irrigated and drought conditions. *Agron. J.*73:

Saleem, M. and Pitman, M.G. 1987. Salinity tolerance of moong bean (*V. radiata* L.) seed production. *Biol. Plantr.* 30: 53-57.

Schneider, S.H. 1989. The changing climate. Sci. Amer. 261: 70-79.

Selvaraj, U., Amnappan, R. S. and Giriclharan, 1986. Co-4 A new high yielding drought tolerant cowpea variety. *Madras Agril. J.* 73: 125-28.

Shadeya, L., Akurdabweni, M., Monti, L. and Fanzo, D. 2000. Detached leaf test as a rapid stress appraisal in cowpea germplasm, *V. unguiculata* (L.) Walp for drought resistance. *Discovery and Innov.* 12: 3-4.

Shakoor, A., Ngugi, E.C.K., Mutohka, H. S. and Kamau, J. W. 1984. Improvement of cowpea varieties for dry areas in Kenya. *East African Agric. and Forestry* J. 44 (Suppl.): 306-11.

Shouse, P. J. 1979. Effect of water deficit on the growth, yield and water use of a field grown cowpea crop. *Dissert. Abst. Intern.*B 40: 1992-1993.

Shouse, P., Jury, W. A., Stolzy, L. H. and Dasberg, S. 1982. Field measurement and modeling of cowpea under water use and yield under stressed and well watered growth conditions. *Hilgardia* 50(6): 1-14.

Singh, B.B., Mai Kodomi, Y. and Terao, T. 1999. A simple screening method for drought tolerance in cowpea. *Indian J. Genet.* 59(2): 211-220.

Singh, B.B., Mai Kodomi, Y. and Terao, T. 1999. Relative drought tolerance of major rainfed crops of the arid tropics. *Indian J. Genet.* 54(4): 437-44.

Singh, G. 1980. Relative salt tolerance of different varieties of guar and cowpea at germination and seedling stages. *Indian J. Agric. Res.* 14: 155-58.

Singh, S.B., Chhabra, R. and Abrol, J.P. 1980. Effect of soil sodicity on the yield and chemical composition of cowpea grown for fodder. *Indian J. Agric. Sci.* 50: 852-56.

Talwar, H.S., Takeda, J., Yashima, S. and Senboku, 1999. Growth and photosynthetic responses of groundnut genetypes to high temperature. *Crop Sci.* 39: 460-66.

Turk, K.J. and Hall, A.E. 1980. Drought adaptation of cowpea II. Influence of drought on plant water status and relations with seed yield. *Agron. J.* 72: 421-27.

Turk, K.J., Hall, A.E. and Asbell, C.W. 1980. Drought adaptation of cowpea. Influence of drought on seed yield. *Agron. J.* 72: 413-20.

Venkateswarlu, B., Maheswari, M. and Saharan, N. 1989. Effect of water deficit on N_2 (C_2H_2) fixation in cowpea and groundnut. *Pl. and Soil* 114: 69-74.

Walker, D. W. and Miller, J.C. Jr. 1982. The relationship between leaf water potential and biological nitrogen fixation in drought resistant and susceptible genotypes of cowpea. *(Vigna unguiculata* (L.) Walp). Abst. *Hort. Sci.* 17(3, II): 503.

Walker, D.W. and Miller, J.C. Jr. 1986. Rate of water loss from detached leaves of drought resistant and susceptible genotypes of cowpea. *Hort. Sci.* 21: 131-32

Warang, M.O.A. and Hall, A.E. 1983. Reproductive response of cowpea to heat stress: genotype differences in tolerance to heat at flowering. *Crop Sci.* 23: 1088.

Warrang, M.O.A. and Hall, A.E. 1984. Reproductive responses of cowpea *(V. unguiculata* (L.) Walp) to heat stress. II. Response to night air temperature *Field Crops Res.* 8: 17-33.

Watanable, S., Hakoyama, S., Terao, T. and Singh, B. B. 1997. Evaluation methods for drought tolerance in cowpea. pp 141-46. In: *Adv. In Cowpea Res.* (Eds. Singh, B. B., Mohan Raj, D.R. Dashiell, K. & Jacki, L.E.N.), IITA and JIRCAS, Ibadan, Nigeria.

Werry, J., Silim, S.N., Knights, E.J., Malhotra, R. S. and Cousin, R. 1994. Screening techniques and source of tolerance to extremes of moisture and air temperature in cool season food legumes. *Euph.* 73: 73-83.

West, D. W. and Francois, L.E. 1982. Effect of salinity on germination, growth and yield of cowpea. *Irriga. Sci.* 3: 169-75.

Yadav, R.B.R. and Patil, B.D. 1984. Screening of cowpea *(V. unguiculata* L.) varieties for drought tolerance, *Seitschrift fur Pflanzenzüchtung* 93(3): 259-63.

CHAPTER – 8

COWPEA AGRONOMY

S. Anitha, E. Sreenivasan and Raj Singh

ABSTRACT

Cowpea is a pulse crop that responds well to scientific management practices. Being a warm weather crop it can withstand a considerable degree of drought and has good potential in dry land farming. It can be cultivated as pure crop and mixed crop. Well-drained loam or slightly heavy soils are best suited. It is generally grown in soils deficient in organic matter. Coupled with this, poor management practices, lack of weeding and timely irrigation etc. have resulted in low and unstable yield in the crop. Organic manures have a significant influence on cowpea productivity. For sustained yield and sound soil health, integrated nutrient management practices involving organic manures, inorganic and biofertilizers can help a lot. Even though, it is a leguminous crop it requires soil application of NPK as a starter dose for better initial establishment. Phosphorus is the most important nutrient required for cowpea. Calcium has equal priority as that of phosphorus especially in acid soils. Foliar application of nutrient was as good or even better than soil application of fertilizers. Among the bio-fertilizers, *Rhizobium* seems to have maximum potential for cowpea. Vesicular Arbuscular Mycorrhizal (VAM) fungus and Phosphate Solubilizing Bacteria (PSB) offers promise in cowpea as it enables the inoculated plants to uptake more effectively the available phosphorus from the soil. Cowpea is mainly grown under rainfed condition. Keeping the field weed free up to 20-25 days after sowing is very important for better yield. Irrigation is the most critical input in summer cowpea followed by weeding and fertilizer application. By adopting improved management practices we can realize yields up to 12-15 quintals of grain per hectare. Being a leguminous crop it also adds N@ of 30-40 kg /ha to the soil and thereby saving N to the tune of about 30 kg/ha to the succeeding crop.

INTRODUCTION

Cowpea commonly known as lobia is one of the principal pulses, which is being grown since ancient times in India. It forms one of the components of the *nava dhanya*, usually used during auspicious ceremonies by Hindus in India. It is a rich source of proteins, calcium and iron. In India cowpea is used as a pulse, a vegetable, a fodder and as a green manure for soil improvement. The crop gives such a heavy vegetative growth and covers the ground so well that it checks the soil erosion in problem areas and can later be ploughed in as green manure. It has considerable promise as an alternative pulse crop in dry land farming.

CLIMATIC REQUIREMENT

Cowpea can be grown in all tropical and subtropical climates. In India it is mainly grown in central and peninsular regions. In northern India it is grown in Utter Pradesh, Punjab, Delhi and Haryana. Being a warm weather crop it can withstand a considerable degree of drought. Under very dry and water stagnated conditions the plant produces a poor crop. But cowpea is more tolerant to heavy rainfall than other pulses. Germination of cowpea is better between 12 and 15°C temperatures. The crop thrives best between 27 and 35°C temperatures. It can also grow under shade of trees but cannot tolerate cold or frost (Chhidda Singh, 1989). Plant height, Leaf area index and pods per plant increased when the crop was sown in warmer day and night temperature with longer day length (Khalfe and Jadhav, 1993).

SOIL

Though cowpea is adapted to a variety of soil types (red loam, black clayey loam, rice fields, black cotton soils, coarse gravel, sandy loam, light sandy soils etc), well-drained loam or slightly heavy soils are best suited to cowpea. It is also grown in poor gravelly sloppy lands in hill tracts and the heavy loamy soils. Some types grown well in the black cotton soils while certain others do best in well-drained sandy loams. In colder climate somewhat light sandy soils are preferred, since on such soils the crop mature earlier (ICAR, 1970). Cowpea requires a neutral soil pH. Saline or alkaline soils are not suited to cowpea cultivation.

CROPPING SYSTEM

Cropping system-involving pulses has been found to be more profitable because of the yield advantage of such system. Pulses besides being rich in protein, they sustain the productivity of cropping system. Agro techniques for such systems need to be developed, rather than for individual crops. It will help in economizing the use of inputs like fertilizers.

Crop Rotations

Cowpea can be grown in rotation with many crops. The most common rotations are as below:

For grain or vegetable purpose:

Cowpea – wheat – mung bean /cheena
Cowpea – potato – urd bean
Maize – wheat – cowpea
Rice – wheat – cowpea
Maize – potato – cowpea
Maize – toria - wheat – cowpea
Rice - rice – cowpea
Rice– cowpea
Rice - mustard – cowpea

For fodder purpose:

Sorghum+cowpea – berseem -maize+cowpea
Maize – berseem -maize+cowpea
Maize– oat -maize+cowpea
Sorghum– berseem -maize+cowpea
Sudangrass– berseem -maize+cowpea
Sorghum+cowpea –oat

Cowpea - berseem -maize+cowpea

Intercropping

Studies have shown that growing cowpea in widely spaced crops and incorporation of their biomass after picking pods can increase soil fertility and yield of companion crop. In this system yield of cowpea is reduced to a great extent as proper spacing and planting geometry is not followed. Experiment evidence have shown that one or two rows of cowpea can be grown in between two rows of pigeon pea, maize, sorghum, etc. Under this system cowpea can yield up to 5-7 q/ha without affecting the yield of main crop. The improvement in this system can further be made by pairing the rows of main crops and by taking one or two rows of cowpea in between two paired rows of main crop. The improvement gained by this system can be glanced from the results of an experiment conducted at Pantnagar (Yadav, 1992). It is clear from the data given in the table 1 that intercropping systems with suitable crops can increase productivity of pulses by about 25 per cent.

Table 1. Yield of pigeon pea and cowpea under intercropping system

Treatment	Yield q/ha	
	Pigeon pea	Cowpea
Pigeon pea pure	12.85	
Pigeon pea + 1 row of cowpea	11.65	3.16
Pigeon pea+ 2 row of cowpea	11.85	3.75

Myaka (1995) reported that sowing cowpea between paired rows of maize increased cowpea yields by 57 per cent compared with in alternating single rows with maize. Kalarani (1995) reported that in terms of LER, gross return and net return, intercropping cowpea in bhindi plots with normal row planting receiving 75 per cent N level was profitable and raising a summer crop resulted in saving of 25 per cent nitrogen for the succeeding rice crop. A trial conducted at Pattambi, Kerala to find out suitable arid legume based intercropping system for the area. Data (Table 2) revealed that in cowpea intercropping had non-significant effect on the yield but get additional yield from intercrops (ICAR, 2002). The highest land equivalent ratio (1.532) was recorded with 4 rows of cowpea +1 row of amaranthus, which was at par with 4+1 (Cowpea : Bhindi), 5+1 (Cowpea: Amaranthus) and 5+1 (Cowpea : Bhindi).

Table 2. Effect of intercrop on yield and LER of cowpea cropping system

Treatments	Yield kg/ha			LER
	Cowpea	Amaranthus	Bhindi	
Sole crop of cowpea	492.16	-	-	1.000
Sole crop of amaranthus	-	2379.71	-	1.000
Sole crop of bhindi	-	-	1575.58	1.000
4 rows of cowpea +1 row of amaranthus	641.58	472.36	-	1.532
4 rows of cowpea +1 row of bhindi	476.33	-	477.91	1.285
5 rows of cowpea +1 row of amaranthus	545.49	418.71	-	1.315
5 rows of cowpea +1 row of bhindi	536.00	-	444.00	1.401
6 rows of cowpea +1 row of amaranthus	439.01	248.51	-	1.027
6 rows of cowpea +1 row of bhindi	429.69	-	280.58	1.085
CD 5%	204.71	-	-	0.2643

FIELD PREPARATION

Cowpea requires a normal seedbed preparation. Field should be prepared by giving two or three cross harrowing and planking. In hard soils one deep ploughing followed by two or three harrowing and planking are sufficient for cowpea. The stubbles and other residues of previous crop should be picked up as far as possible to have a clean and smooth seedbed. Soil should be well leveled to avoid even temporary water logging. For summer season crop a pre-irrigation immediately after harvesting of *rabi* crop is required followed by two to three harrowing. Mulching of field with straw or grass is practiced in cowpea growing countries. Ike (1989) suggested that disc-harrowing method should be adopted for large-scale production of cowpea in a sandy loam soils but if chemical weed control could be applied cheaply, no tillage method would be preferable. Osuji (1990) observed that conventional tillage (ploughing + harrowing) and reduced tillage (ploughing only) practices increased the seedling emergence and overall seedling performance of cowpea compared to no tillage. Addition of grass mulch increased the soil moisture in the root zone and significantly decreased maximum soil temperature and diurnal fluctuations in temperature. This provides a stable environment for seedling establishment and growth than the unmulched soil. Hulugalle and Palada (1990) recommended a combination of minimum tillage and straw mulch as the least risky and hence, the most appropriate soil management system for dry season cowpea production in rice fallows. Mulching increased the soil moisture in the 0-15 cm layer and increased the seed yield of cowpea over no mulching in a sandy clay loam soil at Hebbal, Bangalore (Channabasavanna, 1993). Gumbs and Lindsay (1993) reported that in a clay loam soils of Trinidad, disc ploughing + rotovating and mulching did not significantly affect the growth and yield of cowpea compared to no tillage and unmulched plots. Lindsay and Gumbs (1993) obtained the same result in a silty clay loam and a freely drained loamy soil. Polthanee *et al.* (2000) also opined that growth and yield of cowpea grown following rice were not significantly affected by tillage and no tillage but mulch application significantly increases the growth and yield. This was due to the better ability of mulched plots in storing soil moisture during the growing seasons.

SOWING TIME

Sowing time is one of the important practices that bring drastic changes in the yield. For cowpea, too early and too late sowing are not suitable. Cowpea can be cultivated in almost all season. It can very well be fitted into multiple cropping systems. Cowpea is sown generally with the onset of monsoon, the period ranging from early June to the end of July. *Rabi* cowpea is planted in October and November in southern India. Summer cowpea should be planted in March. For fodder purpose, it may be sown in the beginning of February (Yadav, 1992). For green manuring, the crop should be sown from middle of June to first week of July. In hills this crop is sown in April-May (Chhidda Singh, 1989).

Though in rain fed areas sowing time is governed by the onset of monsoon, the optimum time of sowing of cowpea can be found out based on the rainfall pattern for different situations. In Bangalore, sowing of cowpea in July gave the highest pod yield (Ram *et al.*, 1989). In Andhra Pradesh, sowing of cowpea at the onset of monsoon gave the highest yield (0.68 t/ha) compared with 10, 20 and 30 days after the onset of monsoon (Kumar *et al.*, 1991). Dhaka *et al.*, (1992) reported a reduction in seed yield from 890 to 90 kg/ha with delay in sowing date in the *kharif* season in Rajasthan. In Maharashtra, Jadhav *et al.*, (1994) reported that seed yield of cowpea decreased from 1.66 to 0.68 t/ha as sowing date was delayed from 1st June to 10th July. In a field trial in the *rabi* season at Maharashtra, sowing of cowpea cv. C-152 in 42nd, 44th, 46th or 48th meteorological weeks had no significant effect on seed yield, seed protein content and yield components (Gandhi *et al.*, 1991). The optimum sowing date of summer cowpea at Utter Pradesh was 10th March. Sowing earlier or later this date decreased the seed yield significantly (Rajput, 1994). Under Kerala conditions cowpea can be grown during any season (KAU, 1996a). The most suitable time for sowing in *kharif* is after the first week of June. It can be grown as a floor crop in coconut gardens and as an inter crop in tapioca during May-September. It can be grown as a pure crop in single crop and double crop rice fallows during *rabi* and summer seasons. In Kerala during *rabi* season, there was no difference in grain yield of cowpea under two sowing periods viz. September and October (KAU, 1996b). Cowpea can be grown in homestead gardens through out the year.

SEED RATE

Variation in seed rate depends on several factors such as method of sowing, availability of soil moisture, seed size of a variety, and the purpose for which cowpea is grown. For grain or vegetable purpose, 20-25 kg/ha seed is required for pure crop. When sown mixed with other crops, the seed rate is reduced proportionately. For green manuring and fodder purpose, seed rate should be increased to 30-35 kg/ha. During summer season, vegetative growth is comparatively less. Therefore, closer spacing with higher seed rate should be followed. Seed rate should be increased to about 30 kg for grain and green pods and 40 kg for fodder and green manuring. Tsigewoin *et al.* (2000) reported that a medium seed rate of 40 kg/ha was significantly superior to 30 and 50 kg of seed/ha in green fodder, dry matter crude protein, etc., of cowpea fodder.

METHOD OF SOWING

Broad casting, line sowing or dibbling of seeds was done depending upon the season and the purpose for which cowpea is grown. During rainy season, cowpea is generally sown as broadcast. Sowing in rows is considered superior over broadcasting. For fodder and green manure crop broadcasting is followed. Cowpea was generally sown in flat beds or on ridges/furrows. The yield of cowpea sown in flat beds at Parbhani, Maharashtra in the *kharif* season was the lowest (2.69 t/ha), as against 2.8 t/ha in the standard ridge/furrow method. The yield was highest (5.01 t/ha) when the seeds were sown in the center of the furrow and the ridges modified after about a month such that the plant stems were placed in the center of the ridges (Lawand *et al.* 1993). Under Kerala conditions during *kharif* season, dibbling of two seeds per hole is adopted (KAU, 1996a) for grain and dual-purpose cowpea. Channels of 30cm broad and 15cm deep are drawn at 2m apart to drain off excess rainwater. If broad casting is adopted, the seeds can be sown broadcast over the field and channels drawn after sowing. During the second crops season (September – December) cowpea can be grown as a fringe crop along the rice field bunds (Fig. 1). Sowing can be done on either side of bunds on the day of transplanting the paddy crop. During summer, cowpea can be sown broadcast in rice fallows immediately after the harvest of paddy crop.

Fig. 1. Fringe cropping of cowpea

SPACING

Sowing in rows should be done at distance of 30-45 cm depending upon the duration of variety and time of sowing. For dwarf and bushy type varieties a closer spacing (30 cm) and for semi spreading type a wider spacing (45 cm) should be adopted. For vegetable type (spreading), a spacing of 45 cm between rows and 15 cm between plants is suitable (KAU, 1996a). The row spacing of 30-45 cm and plant-to-plant distance of 8-10 cm may be maintained for *kharif* planting. The row spacing in spring and summer planting should be kept 25-30 cm, as the plants do not grow much (Chhidda Singh, 1989; Yadav, 1992). A spacing of 30 x 15 cm gave higher yield than did spacing of 40 x15 cm and 20 x15 cm (Mahaldar *et al.* 1991; Mali and Mali 1991; Daljith *et al.* 1992 and KAU 2002). Patil *et al.* (1991) and Angne *et al.* (1993) reported significantly higher grain yield of cowpea at 45 cm row spacing. Closer spacing (30 cm) proved significantly superior to wider spacing (45 cm) in respect of seed yield of cowpea at Bangalore and Pattambi (Fig. 2). At Hisar wider spacing proved significantly superior while at SK Nagar,

Gujarat non-significant variation in seed yield was observed. On the basis of the mean data (Table 3), 30 cm row spacing increased the seed yield by 14.25 per cent over 45 cm row spacing (ICAR, 2002).

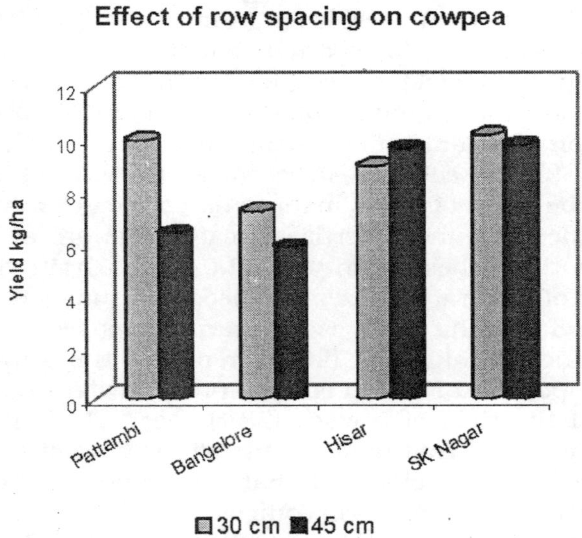

Fig. 2. Effect of row spacing on cowpea

NUTRIENT MANAGEMENT

One of the reasons of low and unstable yield of pulses is no or low use of manures and fertilizers. Pulse growing areas are mostly deficient in soil organic matter and for sustained yield and sound soil health organic manures must be used over inorganic fertilizers (ICAR, 2001; Chaturvedi and Masood Ali, 2002). Organic manures have a significant influence on cowpea productivity. A lot of data have been generated on long term studies in different projects, including dry land projects on integrated use of nutrients proven the utility of FYM, compost, crop residues, bio-fertilizers etc. in pulses.

Organic manures

In poor soils, FYM at 5 –20 t / ha may be given as basal along with first ploughing (Tomy, 1985). Long-term use of fertilizers on soil fertility and yield of cowpea in irrigated Inceptisol of Tamil Nadu showed that the optimum dose of NPK fertilizer in conjunction with FYM produced the highest grain and straw yield over fertilizer alone (Muthuswamy *et al.* 1990). Cowdung and poultry manure improved growth and yield of cowpea compared to inorganic fertilizers (Ahmed *et al.*, 1991). Karmegam and Daniel (2000) reported that in cowpea fresh weight, dry weight and yield were greatest in soils amended with vermicompost. Muthukumar and Udaiyan (2000) observed that application of organic amendments increased the plant growth and AM fungal colonies in cowpea. Oliveira *et al.* (2001) studied the effect of cattle manure on cowpea observed that in the absence of mineral fertilizer, cattle manure at 25 kg/ha produced the highest grain yield (2.0 t/ha). Grain yield in the presence of mineral fertilizer reached the estimated maximum value of 3.03 t/ha with cattle manure applied at 21 t/ha. Sailaja and Ushakumari (2001) reported the superiority of vermicompost and FYM on cowpea productivity. Vermicompost increased the yield of cowpea (23%) over FYM (Anitha *et al.*, 2001a). In a field trail to study the effect of humic substances on cowpea, it was observed that application of humic substances (HS), FYM and vermicompost significantly increase the yield of cowpea compared to control (KAU, 2002). The data presented in table 3 revealed that application of FYM @ 20 t ha^{-1} recorded the highest yield (957 kg ha^{-1}) followed by vermicompost @ 10 t ha^{-1} (835 kg ha^{-1}). Substituting the bulky and costly organic manures like FYM and vermicompost with humic substances reduced the cost of cultivation of cowpea without affecting the yield of cowpea.

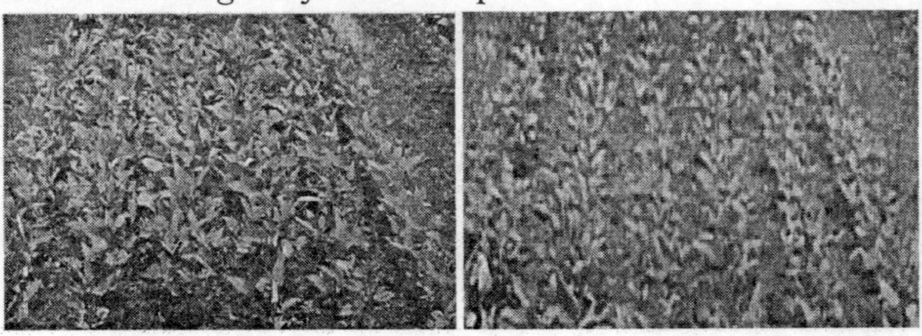

Humic substance applied plot *Control plot*

Fig. 3. Growth of cowpea as affected by humic substances

Table 3. Effect of FYM, vermicompost and humic substances on yield of cowpea.

Treatments	Yield Kg / ha (Pooled data)	Economics	
		Net return (Rs.)	BC ratio
T1- Control	456	1330	1.17
T2- soil application of HS granules @12.5 kg ha^{-1}	635	4205	1.49
T3- seed treatment with HS 6% (5 ml per kg of seed)	643	5015	1.63
T4- 2 sprays of 6% HS at veg. and at flow. stage (2ml/litre)	591	3641	1.44
T5- T2+T4	652	4162	1.46
T6- T3+T4	687	5520	1.67
T7-Vermicompost 10 t ha^{-1}	835	-11087	0.60
T8-Vermicompost 5 t ha^{-1}	691	-3968	0.77
T9-FYM 20 t ha^{-1}	957	-8654	0.68
T10-FYM 10 t ha^{-1}	654	-4701	0.73
CD at 5%	119.63	—	—

Fertilizers

Being a leguminous crop it needs a small quantity of nitrogen (15-20 kg/ha) as a starter dose for early growth period on those soils that are poor in organic matter. The application of fertilizer nitrogen can be adjusted according to organic matter content of the soils (Fig. 3). If the organic matter content of the soil is 0.5 per cent or less full dose of fertilizer nitrogen may be given. If it is between 0.5 per cent and 2 per cent the dose of fertilizer nitrogen can be modified accordingly. If the organic matter content of the soil is above 2 per cent the application of fertilizer nitrogen can be dispensed with (Tomy, 1985).

In the cultivation of cowpea, application of phosphorus is very important as it promotes proper development of roots and function of *Rhizobium* bacteria. It is advisable to apply phosphatic fertilizers according to the soil test recommendations. Apply 50-60 kg of P_2O_5 per hectare in the soil before sowing in case soil test values are not available. Sandhu *et al.* (1976) found that cowpea favourably responded to phosphorus application up to 25 kg P_2O_5. At Delhi, cowpea responded significantly to P application. However, increase in the level of P beyond 20 kg P_2O_5/ha did not increase the yield (Ahlawat *et al.*, 1979). Application of 60 kg P_2O_5/ha increased the grain yield by 6.9 and 49.4 per cent over 30 kg P_2O_5/ha and no phosphorus, respectively. However, there was no marked response beyond 30 kg P_2O_5/ha (Sekhawat, 1988). Cowpea response to phosphorus was up to 40-100 kg/ ha [Singh et al. (1989); Amin *et al.* (1990); Mahaldar *et al.* (1991); Naceur and Ben (1991); Sawant *et al.* (1992); Kher *et al.* (1994) and Pitan *et al.* (2000)]. Phosphorus application improved the nodulation, plant uptake of N, P and K and seed yield (Mali and Mali, 1991) and increased growth and yield attributes of cowpea (Abdel, 2000; Kolawole *et al.*, 2000).

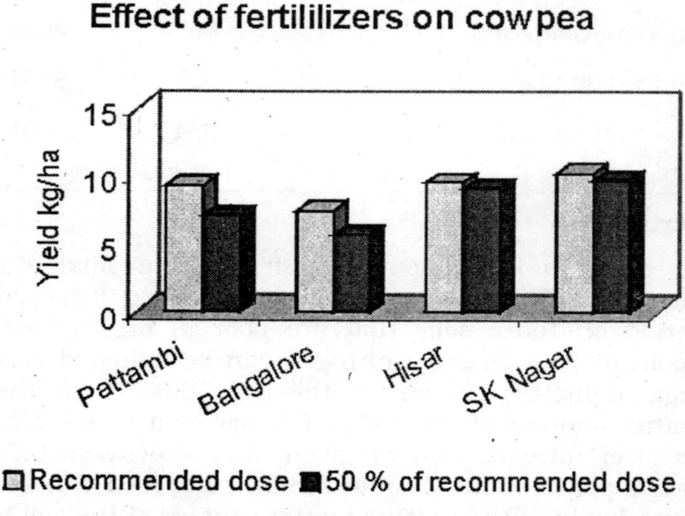

Fig. 4. Effect of fertilizers on the yield of cowpea

Raj and Patel (1991) reported the highest seed yield with 20 kg N and 80 kg P_2O_5/ha. Field response of cowpea up to 50 kg N and 75 kg P_2O_5/ha was reported by Singh (1991). The data presented in Table 5 revealed that application of recommended dose of fertilizers (20 kg N + 40 kg P_2O_5 / ha) proved significantly superior to 50 per cent of recommended dose (10 kg N + 20 kg P_2O_5/ha) at Pattambi, Bangalore and SK Nagar, while at Hisar non- significant variation was observed. On the basis of the mean data, recommended dose of fertilizers increased the seed yield (15.88%) over 50 per cent of recommended dose (Fig. 4).

Increase in pod yield of cowpea with increased levels of P_2O_5 and K_2O up to 80 kg / ha was observed by Pushpendra *et al.*, (1990). Amin *et al.* (1990) reported the highest grain yield of cowpea with N, P_2O_5 and K_2O at 25: 60: 40 kg/ ha. Murthy *et al.* (1990) reported that applying N, P_2O_5 and K_2O at 25: 25: 25 kg/ ha gave a yield of 1.06 t/ha compared with 0.79 t/ha without NPK. Mahadurappa and Sharanappa (1990) observed an increase in grain yield with increased dose of fertilizer and NPK @ 25:50:25 kg / ha gave the highest yield over 50 per cent of the above dose and no fertilizers. Cowpea cv. Kurutholapayar was given 0-30 kg N + 0-60 Kg P_2O_5 + 0-30 kg K_2O/ha at Kumarakom, Kerala. Green pod yield increased with up to 10 kg N + 20 Kg P_2O_5 + 10 kg K_2O / ha (Syriac and Nair, 1994). Application of potassium fertilizer promoted the growth of cowpea and mitigates the impact of water stress in plants when subjected to sub-optimal soil stress (Sangakkara *et al.*, 2001).

Calcium has equal priority as that of phosphorus in cowpea, calcium may be supplied as calcium sulphate (gypsum) in alkaline soils and as calcium carbonate in acid soils (@ 250–400 kg/ha) may be applied as basal for cowpea. Adhikari *et al.* (1989) reported increased seed yield of cowpea with incremental dose of lime.

Crop removal of sulphur is more in pulses and oil seeds. In fact, for pulses, S is the second most important nutrient after P and it has been found to interact with P, as both are anions. Singh *et al.* (1991) opined that sulphur application has improved the grain yield and quality of legume crops. Bansal (1991) noticed increased seed yield, N, P, K and S content and uptake in cowpea with increased rate of sulphur application. Aulakh and Chhibba (1992) reported that pulses respond well to sulphur application in S-deficient soils. Cowpea response to sulphur application on sulphur deficient soils were reported by

Rattan *et al.* (1995). Gracy *et al.* (1998) studied the effect of graded doses of phosphorus and sulphur on the yield and nutrient uptake of rain fed cowpea in Kerala. Maximum grain yield 645 kg/ha and nutrient uptake were obtained with application of 60 kg P_2O_5/ha and application of sulphur did not show any remarkable influence on growth, yield and nutrient uptake of cowpea. This can be attributed to the comparatively higher status of available sulphur in Kerala soil. Patel *et al.* 2002a reported similar results in Gujarat. Shankaralingappa *et al.* (2002) studied the interaction effect of phosphorus and sulphur on cowpea on a sulphur deficient, high P, red sandy clay loam soils of Karnataka Combined application of 50 kg P_2O_5 and 20 kg S /ha produced significantly higher grain yield (1757 kg/ ha). A significant increase grain protein content (24.58%) and methionine content (Shankaralingappa *et al.*, 2000a) and a significant synergistic effect on the uptake of N, P and S (Shankaralingappa *et al.*, 2000b) were observed.

Method of fertilizer application

Lime may be applied at the time of first ploughing. Half the quantity of N, whole phosphorus and potash is applied at the time of final ploughing. The remaining N may be applied 15 –20 days after sowing. Fertilizer should be applied by drilling at the time of sowing in such a way that it is placed at about 5-7 cm below the seed. Stamford *et al.* (1990) reported that band application and broadcast of NPK fertilizers had no significant influence on cowpea grain and dry matter yield. Srinivasan and Ramasamy (1992) reported that spraying of 2 per cent DAP at 20 and 30 days after sowing produced similar yield to soil application of 12.5 kg N and 25 kg P_2O_5 and higher yield than 2 per cent urea sprays. Cowpea seed yield can be improved substantially by topdressing with moderate rate of P provided that P deficiency is identified significantly earlier (Ikombo *et al.*, 1993). Foliar application of nutrient was as good or even better than soil application of full dose of recommended fertilizer. Foliar application of 2 per cent urea recorded the highest grain yield of cowpea compared with soil application of fertilizers. It may be attributed to the better utilization of nutrients applied through foliage and due to the uniform growth of plants receiving foliar nutrition (KAU, 1996b). Soil application of recommended N and P_2O_5 + 2 per cent DAP spray twice at flowering and second at 15 days after first spray found to record the highest seed yield (1134 kg/ ha), net income (3471/ ha) and BC ratio (1.96) (Parasuraman, 2001).

MICRONUTRIENTS

Deficiencies of boron (B), copper (CU), iron (Fe), manganese (Mn), molybdenum (Mo) and zinc (Zn) have been widely reported in pulses. Correction of micronutrient deficiency is warranted for improving the yield potential of pulses. Among the micronutrients Zn, Fe, B, Mn and Mo improve the yield appreciably and foliar spray of micronutrients proves to be economical in pulses. Mundra and Bhati (1991) reported increase in seed yield of cowpea with increased rate of $MnSO_4$ application. $MnSO_4$ application increased total N, P, Fe and Mn uptake. Agbenin *et al.* (1990) reported a yield depression in cowpea with applied B under field condition. Chowdary *et al.* (2000) observed a lack of response to applied B in soils where wide spread deficiency has been reported strongly may be attributed to low B requirement of cowpea. Savithri *et al.* (2001) reported that application of Mo (@ 0.5 kg/ha) was found to be economical and optimum for enhancing the productivity of cowpea. Foliar spray of 0.05 to 0.1 per cent Na_2MoO_4 is not only economical but also effective. Effect of micronutrient on the productivity of cowpea was studied at Bangalore, Pattambi and S. K. Nagar under the All India Co-ordinated Research Project on Arid Legumes (ICAR, 2002). The data presented in Table 4 showed that at Bangalore non-significant variation in seed yield was observed on account of foliar spray of all the micronutrients. While at Pattambi, 0.5% $FeSO_4$ + 0.5% $ZnSO_4$ spray at 45DAS proved most effective and increased the seed yield by 44.9 per cent as compared to control. Where as at S. K. Nagar, 0.5% $FeSO_4$ + 0.5% $ZnSO_4$ spray both at 25 and 45 DAS proved most effective and increased the seed yield by 27.7 per cent when compared with control (Table 4).

Table 4. Effect of iron and zinc sprays on yield of cowpea

Treatments	Yield q/ha		
	Bangalore	Pattambi	S.K. Nagar
T1- Control	16.73	4.24	16.36
T2- 0.5% $FeSO_4$ spray at 25DAS	15.87	4.47	17.40
T3- 0.5% $FeSO_4$ spray at 45DAS	13.13	4.69	18.05
T4- T2+T3	13.61	4.75	18.45
T5- 0.5% $ZnSO_4$ spray at 25DAS	14.81	4.35	18.31
T6- 0.5% $ZnSO_4$ spray at 45DAS	14.40	4.69	18.17
T7- T5 +T6	15.11	4.75	19.10
T8- T2+T5	12.94	5.26	18.63
T9- T3 +T6	12.71	6.16	18.09
T10- T4 +T7	13.76	3.69	20.89
CD 5%	NS	0.64	1.71

BIOFERTILIZERS

Among the bio fertilizers, *Rhizobium* seems to have maximum potential for cowpea. *Rhizobium* induces better root nodulation and there by reduces the requirement of nitrogen fertilizer. Saxena *et al.* (1971) reported increased yields of cowpea varieties grown in summer seasons at Pantnagar due to *Rhizobium* inoculation (Table 5). Cowpea cv. Pusa Phalguni when inoculated with *Rhizobium* significantly increased the yield, shoot dry weight and leghaemoglobin content of nodules (Sivaprasad and Shivappashetty, 1980).

Table 5. Effect of *Rhizobium* inoculation on the grain yield (q/ha) of three cowpea varieties

Variety	Uninoculated	Inoculated	Mean
Pusa Phalguni	4.66	6.66	5.78
Pusa Dophasali	5.77	6.66	6.22
F. S. 68	7.33	7.77	7.55

Sekhawat (1988) conducted field experiment at Jaipur found that seed inoculation with *Rhizobium* produced significantly higher grain yield (6.58 Q/ha) over uninoculated (1.58 Q/ha). This result is in conformity with Singh *et al.* 1989. Vimal *et al.* (1990) observed that the pH range most suitable for the growth of *Rhizobium* is between 6.5-7.0. Raj and Patel (1991) reported that *Rhizobium* inoculation increased the grain yield by 0.15 t/ha and fodder yield by 0.65 t/ha. In an experiment to study the nitrogen requirement of cowpea in different soil types, Lindsay and Gumbs (1993) reported that *Rhizobium* inoculation gave significantly higher yield than 0 and 20 kg N in a clay loam soil. However, the highest yield was obtained by 40 kg N. On loamy soil, seed yield was decreased by 40 kg N/ha and increased by *Rhizobium.* Seed inoculation with *Rhizobium* gave a seed yield of 1.19 t/ha compared with 0.95 t/ha without seed inoculation for cowpea cv. Pusa Komal in Utter Pradesh (Rajput, 1994).

Vesicular Arbuscular Mycorrhizal (VAM) fungus offers promise in cowpea as it enables the inoculated plants to uptake more effectively the available phosphorus from the soil. There is also an enhanced absorption of water, N, K and micronutrients in inoculated plants. Bagyaraj and Manjunath (1980) reported an increased concentration of phosphorus and zinc in mycorrhiza-inoculated cowpea. Increase in plant height, total dry matter of root, shoot and grain were reported by Chhabra *et al.* (1990) in cowpea inoculated with VA mycorrhiza and there was a significant increase in N, P and K content of inoculated plants. Ikombo *et al.* (1991) observed an increased P absorption in VAM inoculated cowpea. Pai *et al.* (1994) found high calcium levels in all parts of cowpea plants inoculated with a VAM fungus *Glomus fasiculatam.* Geetha-kumari *et al.* (1994) reported that the recommended level of 30 kg P_2O_5/ha can be reduced to 22.5 kg for grain cowpea by combined application of mussorie rock phosphate and mycorrhiza. Muthukumar and Udaiyan (2000) reported that mycorrhizal fungi act as a strong sink for photosynthates and reduced the nutrient imbalances in cowpea. VAM inoculation increased plant height, number of branches per plant, pods per plant, dry matter production, 100 seed weight and seed yield of cowpea (Abdel, 2000).

Phosphate solubilizing bacteria (PSB) are another group of bio fertilizers for increasing phosphorus availability to cowpea. Desai *et al.* (2001) reported a significantly higher yield of cowpea treated with cultures of phosphate solubilizing microorganism in comparison to the recommended dose of P_2O_5.

Dual application of efficient strains *Rhizobium* and VAM-fungi coupled with the application of FYM is found beneficial in cowpea. Ramaraj *et al.* (1990) and Thyagarajan *et al.* (1992) reported that the dual application of cowpea with VAM fungus and *Rhizobium* increased the pod yield, rhizobial nodules, mycorrhizal colonization, biomass, shoot N and P and yield to a great extent than when either was inoculated alone. Thyagarajan and Ahmed (1993) from their experiment observed that co-inoculating with a selected strain of VAM fungus could enhance the effectiveness of *Rhizobium.* Soddi *et al.* (1994) and Beena (1999) reported that 50 per cent of N and P recommended fertilizer could be reduced by seed treatment of *Rhizobium* and soil inoculation with *Glomus fasiculatam* in cowpea.

METHOD OF APPLICATION OF BIOFERTILIZERS

Soil application

Soil application is recommended generally for all types of biofertilizers except *Rhizobium.* The recommended practice is to apply 1-2 kg of the biofertilizers after mixing with FYM or compost in the ratio 1:25. The biofertilizers can be applied at the time of sowing or intercultural operations.

Seed treatment

On an average 250–500 g of commercially available biofertilizers will be required for treatment of seeds meant for one-hectare area. Thick slurry of the biofertilizers is initially prepared by mixing it with water. Instead of water, 10 per cent jagerry solution or 5 per cent sugar solution containing 40 per cent gum arabic or thick rice gruel of previous day for mixing the culture can also be used, in order to ensure better stickiness of the inoculants with the treated seed material. The seed material is gently mixed with this slurry, dried in shade and sown immediately. Any damage to the seed coat should be avoided. Do not mix directly either the culture or inoculated seeds with chemical fertilizers.

In Kerala, the soils are acidic in nature and hence lime pelletting is found to be beneficial along with *Rhizobium* inoculation. Add finely powdered (300 mesh) calcium carbonate to moist freshly *Rhizobium* treated seeds and mix for 1- 3 minutes until each seed is uniformly pelletted. Depending upon the seed size, the following quantity of lime will be required. For small seed (1kg lime/ 10 kg of seed), medium sized seed (0.6 kg lime / 10 kg of seed) and large sized seed (0.5 kg lime/ 10 kg of seed). Spread out the pelletted seeds on a clean paper to harden. Sow them as soon as possible. However, lime pelletted seeds can be stored up to one week in a cool place prior to sowing. Lime coating is required only for seeds that are to be sown in acid soils. Pelletting with ordinary agricultural lime is not good because of its larger particle size. On no account hydrated lime should be used for pelletting. The dry pellet should be firm enough to resist moderate pressure. It should appear dry without loose lime on its surface or in the container. The lime pelletted seeds can be mixed with fertilizer and sown. However the period of contact between fertilizer and the pelletted seed should be as short as possible. Pelletted seeds should not be sown into a dry seedbed (KAU, 1996a).

AFTER CULTIVATION

Hoeing at the time of application of second dose of N will help to give adequate aeration to the soil and help the root system to spread easily. For grain and dual-purpose cowpea decapitation is found to be advantageous as the crop shows trailing tendency (KAU, 1996a)

WEED MANAGEMENT

Heavy weed growth during the early stages of the crop is a major constraint limiting the productivity of cowpea. Weed growth reduces the crop yield by 50-62 per cent over weed free situation. Weed free condition through out the crop growth period recorded the highest grain yield in cowpea as a result of competition free environment. Manual weeding is impractical due to high cost and non-availability of labourers. Therefore, an integrated approach of appropriate agronomic practices like, optimum sowing time, proper cropping geometry, optimum plant density, intercropping, timely manual / mechanical/ chemical weeding and inter-cultivation etc. is essential to minimize the weed infestation. Effective control of weeds in the first 20-25 days of the crop season is essential. One hand weeding/hoeing may take care of most of the weeds in cowpea. Later on, if required, it may be repeated after 20-25 days after first weeding. For chemical weed control, basalin 1 kg a.i./ha in 800-1000 litres of water as pre-planting spray (Chhidda Singh, 1989) or lasso at 1 litre a.i./ha as pre-emergent spray (Yadav, 1992) can control the weeds to a great extent. Hoeing and hand weeding at 14 and 28 days after emergence of cowpea reduced the weed growth and resulted in the highest grain yield (Elliot and Moody, 1992). Gracy *et al.* (1995) reported that Pendimethalin 0.75 kg a.i./ha combined with hand weeding at 35 days after sowing recorded the highest weed control efficiency and a two-fold increase in marginal benefit cost ratio in integrated weed management over normal weeding (Table 6). Mirza *et al.* (2001) opined that Fluchloralin 2 kg a.i./ha + mulching paddy straw 10 t/ha for 40 days recorded significantly higher seed yield (1766 kg/ha). Patel *et al.* (2002b) also reported that Pendimethalin 0.75 a.i. kg/ha combined with hand weeding at 35 days after sowing gave the higher seed yield and net realization.

Table 6. Effect of weed management practices on weed index, seed yield and BC ratio in cowpea

Treatment	Weed index	Seed yield kg/ ha	Benefit cost ratio
Weedy check	62.2	232	1.10
Pendimethalin 0.75 kg /ha	46.8	326	1.22
Pendimethalin 1 kg /ha	35.2	397	1.39
Pendimethalin 1.25 kg /ha	38.2	378	1.25
Hand weeding 20DAS	42.3	354	1.15
Hoeing 20DAS	35.9	393	1.43
Pendimethalin 0.75 kg /ha +Hand weeding 35DAS	18.1	502	1.69
Pendimethalin 0.75 kg /ha + Hoeing 35 DAS	38.8	375	1.32
Weed free		613	1.45
CD at 5%		85	

WATER MANAGEMENT

Cowpea is mainly grown under rain fed condition. The rainy season crop requires no irrigation but proper care should be taken to drain excess water from the field. Gumbs and Lindsay (1993) investigated the effect of water logging in cowpea grown in silty clay loam fond that flooding the soil at the vegetative stage for 1, 2,4 or 6 days reduced seed weight / plant and several vegetative plant growth parameters. The reductions were greater for longer water stagnations. Flooding for 1 or 2 days at the flowering and pod setting stages did not affect any of the plant parameters measured (plant height, leaf number, leaf area, leaf, stem and root dry matter). However, flooding for more than 2 days at the flowering stage reduced the plant parameters. If the soil was allowed to drain for 2 days after being flooded for 2, 4 and 6 days, flooding were not as damaging as continuous flooding. The early sown rainy season crop may need one or two irrigation in the pre-monsoon period. For raising summer crop, five to six irrigations may be given. The number and frequency of irrigation depends upon the soil type and weather prevailing during the growth period. Generally the crop should get irrigation at an interval of 10-15 days.

In early *kharif* sown cowpea in Karnataka, application of 5 cm water/irrigation during two dry spells at 27 and 55 days after sowing gave seed yield of 1.15 t/ ha compared with 0.68t under rain fed condition (Murthy *et al.*, 1990). Irrigation was the most important input followed by weeding and fertilizer application in summer cowpea (Joy *et al.*, 1991). Experiments conducted at Pantnagar in variety Pusa Phalguni showed that irrigation significantly increased the grain yield of cowpea over without irrigation. Increasing moisture supply from dry to medium wet regime resulted in significant increase in yield; however, increase beyond medium wet regime did not improve yield (Yadav, 1992). Sawant *et al.* (1992) observed higher seed yield of cowpea when irrigation given at flowering and pod development stage. Gumbs and Lindsay (1993) reported that irrigation at 8 days interval from the flowering stage was as good as irrigation at 4 or 8 days interval through out the crop. Water management studies carried out at Agronomic Research Station, Chalakkudy, revealed that cowpea cv. Kanakamani irrigated at IW/CPE at 0.25, 0.5, 0.75, 1 and irrigated at branching + flowering + pod formation stage gave mean seed yield of 1063,1084,1267,1302 and 1231 kg/ ha respectively. For optimum growth and yield of cowpea requires irrigation at IW/CPE of 0.75 (Santhakumari *et al.*, 1994). Summer cowpea at Gujarat requires irrigation at 0.8 IW/ CPE ratio to produce more number of pods per plant, length of pods, test weight and seed yield (973 kg/ ha) and this was superior to 0.4, 0.6 IW/ CPE ratios (Kher *et al.*, 1994). Manjunatha *et al.* (2000) observed a 27 per cent higher grain yield, a significant water saving (21%) and higher BC ratio (1.54) for micro sprinkler than the surface method of irrigation. Anitha *et al.* (2002) reported that irrigation schedule had a significant effect on the seed yield of summer cowpea (Table 7). The irrigation at 0.8 IW/CPE recorded significantly higher yield (936 kg ha^{-1}) and WUE (27.10 kg/ha/cm) compared to irrigation at 0.4 IW/CPE (795 kg ha^{-1}). This increased seed yield at 0.8 IW/CPE ratios may be attributed to timely and adequate moisture availability, which helped in proper utilization of nutrients and also better formulation and accumulation of photosynthates. Cowpea without water stress recorded four times higher yield (869 kg/ ha) than rain fed crop which recorded 219 kg/ ha. Irrigation at flowering stage alone increased the seed yield by 194 per cent than rain fed treatment while irrigation at vegetative stage recorded 175 per cent yield increase and corresponding WUE were 16.0 and 14.5 kg/ha/cm. The response to irrigation is in the order of flowering>pod filling>vegetative stage (TNAU, 2002).

Table 7. Effect of irrigation on dry matter production, yield and WUE of cowpea

Treatments	Dry matter production kg/ha (Pooled)	Grain yield kg / ha (Pooled)	WUE kg/ha/cm
IW/CPE at 0.4	1778	795	26.94
IW/CPE at 0.8	1838	936	27.10
CD 5%	NS	46.67	

ABIOTIC STRESS MANAGEMENT

Cowpea is grown in a wide range of environment, and water stress is the principal environmental factor limiting the growth and yield. Other important stresses, which tend to be more localized, include salinity, water logging, mineral toxicity and deficiency. These stresses individually or in combination invariably produce adverse effects through water scarcity. Management of abiotic stresses may help in enhanced and sustained production of cowpea. The sensitivity of cowpea to moisture stress varies widely at different developmental stages but usually most sensitive to drought at the onset of flowering and during the reproductive phase. Water stress during reproductive stage significantly reduced the number of pods per plant and the WUE was significantly reduced by water stress during vegetative and reproductive stages in cowpea (Ferreira *et al.*, 1991). Carvalho *et al.* (2000) reported that the highest sensitivity of cowpea to water deficit was during the vegetative phase followed by flowering and fruiting phase. Different levels of water deficit in different phases of crop growth cycle of cowpea caused an yield reduction of 35-100 per cent in the vegetative phase, 34-81 per cent in the flowering phase and 32-73 per cent in the fruiting stage.

Maintenance of adequate potassium in the soil improves plant water relations, photosynthesis and yield. Sangakkara *et al.*, (2000), reported beneficial effect of potassium in over coming soil moisture stress and increasing physiological parameters and carbon partitioning in cowpea. Application of potassium fertilizer promoted the growth of cowpea and mitigates the impact of water stress in plants when subjected to sub-optimal soil stress (Sangakkara *et al.*, 2001).

Similarly seed treatment with thiourea followed by foliar sprays at vegetative and pre-flowering stages significantly mitigated the detrimental effect of field drought through enhancement of photosynthetic efficiency and nitrogen metabolism and get higher yield of cowpea (Table 8). Anitha *et al.* (2001b) and Yadav *et al.* (2002) reported 15 and 25.6 per cent yield increase respectively in grain cowpea with seed soaking with thiourea followed by two foliar sprays one at vegetative and one at flowering stages. The data given in table 10 again revealed the superiority of seed soaking of thiourea with two foliar sprays one at vegetative and one at flowering stages on cowpea productivity (KAU, 2002).

Table 8. Effect of thiourea application on cowpea

Treatment	Yield kg / ha		
	2000	2001	Pooled
Control	875.31	442.92	659.11
500 ppm TU seed soaking	886.87	477.80	682.34
Water soaking	950.76	459.22	704.99
500 ppm TU spray one (veg. Stage)	1069.63	532.33	800.98
500 ppm TU spray one (flow. Stage)	825.93	509.14	667.78
500 ppm TU 2 sprays (veg. and flow. stage)	893.08	497.52	695.30
TU seed soaking + 1spray (veg. Stage)	893.24	486.43	689.84
TU seed soaking +1 spray (flow. Stage)	947.66	481.29	715.47
TU seed soaking + TU 2 sprays (veg. and flow. stage)	1072.14	534.33	803.25
Water soaking+ 1spray (veg. Stage)	941.06	498.08	719.57
Water soaking+ 1 spray (flow. Stage)	847.36	474.21	660.78
Water soaking+ TU 2 sprays (veg. and flow. stage)	1058.12	517.76	787.94
CD 5%	98.11	49.23	75.18

HARVESTING AND THRESHING

Green pods for use as vegetable can be harvested 45 to 90 days after sowing depending on the variety. Pods should be harvested while tender, otherwise yield may be poor and pods may develop fibers due to longer retention on the plants. For grains, the crop can be harvested in about 90 to 125 days after sowing when pods are fully matured. In some varieties all pods do not mature at the same time and if left too long on the plant after maturity they are likely to shatter. Therefore, these are hand picked 2 or 3 times and stored. In other varieties the plants are pulled out when most of the pods are mature. The crop should be then dried on a threshing floor tread by cattle or beaten with sticks, and winnowed. The threshed grains should be dried in sun before storage. Harvesting interval has significant influence on the seed yield and seed quality of cowpea. Srimathi *et al.* (1999) reported the highest seed yield (932.2 kg/ha) in plants harvested at 20 days interval after 75 DAS, followed by 10 days interval (916.4 kg/ha). Seeds collected from these treatments recorded highest seed germination (96%) and vigour index of seedling. For fodder, the cutting of the crop depends upon the need and the stage of growth of the component crop grown with it. In general, the crop should be cut when it attains the age of 40-45 days. When cowpea is grown as *in situ* green manure crop, it should be incorporated at the flowering stage.

Yield

A good crop of cowpea yields about 12-15 quintals of grain and 50-60 quintals of straw per hectare. If the crop is raised for fodder purpose, 250-350 quintals of green fodder is obtained from one hectare of pure crop.

REFERENCE

Abdel Ati, Y. Y. 2000. *Assiut J. Agri. Sci. Egypt.* 31 (2) : 21-38

Adhikari, R. R., Timsina, J. and Tripathi, B. P. 1989. *Tropical Grain Legume Bulletin.* **36** pp 36-37

Agbenin, J. O., Lombin, G. and Owonubi, J. J. 1990. *Fertilizer Research,* **22** (2): 71-78.

Ahalawat, I. P. S., Saraf, C. S. and Singh, A. 1979. *Indian J. Agron.* **24** : 247-249.

Ahmed, S. S., Kandil, M. M. and Al Ansi, N. A. 1991. *Annals of Agri. Sci. Moshtohor.* 29 (3) : 1215-1220

Angne, M. N., Patil, R. A., Mahadkar, U. V. and Khanvilkar, S. A. 1993. *J. Maharashtra Agrl. Universities.* 18(1) : 121-122

Anitha, S., Prema, A., Helen, S. and Raji, P. 2001a. *11th Swadeshi Science Congress,* 6th to 8th November 2001 KFRI, Peechi. pp.9

Anitha, S., Purushothaman, S.M. and Sreenivasan, E. 2001b. National Symp. on Pulses and Oilseeds for Sustainable Agriculture at TNAU,,Coimbatore, 29th to 31st July. pp. 102

Anitha, S., Gracy Mathew, Sreenivasan E. and Purushothaman, S.M. 2002. First National Symposium on Arid legumes for Food, Nutritional Security and promotion of Trade, CCS Hisar 15th to16[th] May pp.158

Amin, M., Abedin, M. Z., Naullah, G. M. P. and Kamal, A. M. A. 1990. *Bangladesh J. Agric. Research.* **15** (1) : 32-37

Aulakh, M. S. and Chhibba, I. M. 1992. *Fert. News* **37** (9) : 33-45

Bagyaraj, D. J. and Manjunath, A.1980. *Pl. Soil.* **55** : 495-498

Bansal, K. N. 1991. *Madras Agric. J.* **78** (5-8) : 188-190

Beena, S. 1999. M. Sc. Thesis. Kerala Agricultural University, Kerala

Carvalho, J. de. A., Perira, G. M., Andrade, M. J. B. and Roque, M. W. 2000. *Ciencia e Agrotecnologia.* 24 (3). 710-717

Channabasavanna, A. S., Havanagi, G. V. and Setty, R. A. 1993. *Karnataka J. Agric. Sci.* **6** (10) : 57-58

Chaturvedi, S. K. and Masood Ali. 2002. *The Hindu Survey of Indian Agriculture 2002.* Kasturi and sons Ltd, kasturi Buildings. Chennai. pp. 63-69

Chhabra, M. L., Singh, R. P., Jalali, B. L. 1990. Proc. National conference on Mycorrhiza, Haryana Agricultural University, Hisar, 14th to 16th February pp.150-151

Chowdary, M. U., Ullah, H., Rahman, A. and Shahidullslam. 2000. *Legume Research,* **23** (1): 9-14

Chhidda Singh, 1989. *Modern Techniques of Raising Field Crops.* Oxford and IBH publishing Co. Pvt. Ltd. New Delhi.pp 237-244

Daljit Singh, Sandhu, K. S., Saimbhi, M. S. and Singh, D. 1992. *J Research panjab Agrl. University* **29** (3). 345-346

Desai, N. C., Navadia, L. J., Mehta, B. P., Intwala, C. G. and Ardeshna, R. B. 2001. National Symp. on Pulses and Oilseeds for Sustainable Agriculture at TNAU, Coimbatore, 29th to 31st July. pp.102

Dhaka, B. R., Poonia, B. L. and Keshwa, G. L. 1992. *Madras Agric. J.* **79** (7) : 412-414

Elliot, P. C. and Moody, K. 1992. *Tropical Pest Management,* **38** (1) : 5-8

Ferreira, L. G. R., Costa, J. O., Albuquerque, I. M. de. and De Albuquerque, I. M. 1991. *Pesquisa Agropecuaria Brasileira* . **26** (7) : 1049-1055

Fageria, N. K. 1991. *Tropical Agriculture* **68** (4) : 384-388

Gandhi, D. V., Wagh, R. G. and Thorat, S. T. 1991. *Agric. Sci. Digest Karnal.* **11**(4) : 178-180

Geethakumari, V.L., Pushpakumari, R., Sheela, K.R. and Sivaprasad, P. 1994. *J. Soil Biol. Ecol.* **14** : 25-28

Gracy Mathew, Sreenivasan, E. and Jose Mathew. 1995. *Indian J. Weed Sci.* 27 (1&2) : 42-44.

Gracy Mathew, Sreenivasan, E. and Anilakumar, K.1998. *J Tropical Agriculture.* **36**: 31-33

Gumbs, F. A. and Lindsay, J. I. 1993. *Tropical Agrl.* 70(1) : 27-32

Hulugalle, N. R. and Palada, M. C. 1990. *Soil and Tillage research,* **17** (1-2) : 101-113

ICAR, 2001. Proceedings of XVII Annual Workshop of AICRP on Arid Legumes, ARS Durgapura, Jaipur, Rajasthan, 14th to 15th May 2001. pp. 1

ICAR,2002. *Annual Progress Report 2001-2002,* All India Coordinated Research Project on Arid Legumes, CAZRI Jodhpur, Rajasthan. pp. 121-130

Ike, I. F. 1989 Nigerian J. Technological Research, 1(1) : 7-10

Ikombo, B. M., Edwards, D. G. and Asher, C. J. 1991. *Australian J. Agri. Reseach.* 42(1): 129-138

Ikombo, B. M., Edwards, D. G., Asher, C. J and. Barrow, N. J. 1993 Proc. Twelfth international Plant Nutrition Colloquium, Perth, Western Australia, 21st to 26th September 1993. pp. 379-382

Jadhav, A. G., Shinde, S. H. and Pol, P. S. 1994. *Indian J. Agron.* **39** (3) : 484-485

Joy, P. P., Bridigit, T. K. and Neema, V. P. 1991. *Legume Research.* **14** (1) : 33-36

Karmegam, N. and Daniel, T. 2000. *Envt. and Ecology.* 18 (2) : 367-370

Kalarani, S. 1995. M. Sc. Thesis. Kerala Agricultural University, Kerala

KAU. 1996a. *Package of Practices Recommendations crops,* Kerala Agricultural University, Thrissur, Kerala. pp 39-42

KAU. 1996b. *Annual Research Report,* Regional Agricultural Research Station, Pattambi, Kerala Agricultural University. Thrissur.

KAU. 2002. *XXIVth Zonal research and Extension Advisory Committee Workshop, Central Zone,* Regional Agricultural Research Station, Pattambi Kerala Agricultural University, Thrissur, Kerala, 30th May 2002. pp 15-17

Khalfe, S. D. and Jadhav, B. B. 1993. *Annals of Plant Physiology.* **7** (2): 175-181

Kher, G. C., Patel, J. C., Patel, B. S. and Malavia, D. D. 1994. *Indian J. Agron.* **39** (1) : 175-177

Kolawole, G. O., Tian GuangLong., Singh, B. B. and Tian, G. l. 2000. *J. Plant Nutrition,* **23** (6) : 731-740

Kumar, P. V., Rao, U. M. B. and Singh, S. P. 1991. *Indian J. Agric. Sci.* **61** (1) : 31-35

Lawand, B. T., Rajput, S. G. and Patil, V. K. 1993. *Madras Agric. J.* **80** (7) : 407-408

Lindsay, J. I. and Gumbs, F. A. 1993. *Tropical Agriculture,* **70** (3) : 214-219

Mahadurappa, K. and Sharanappa. 1990. *Current research.* University of Agric. Sci. Bangalore, **19** (10) : 172-173

Mahaldar, S. R., Wagh, R. G. and Patil B. P. 1991. *J. Maharashtra Agrl. Universities.* 16 (1) : 100-101

Mali, O. P. and Mali, A. L. 1991. *Indian J. Agric. Sci.* **61** (9) : 672-673

Manjunatha, M. V., Shukla, K. N. and Chauhan, H. S. 2000. *6th International micro irrigation Congress 2000,Cape Town,South Africa, 22nd to 27th October.* pp.1-8.

Mirza karim Baig,. Nanjappa, H. V., Ramachandrappa, B. K. and Chandrakumar, S. S. 2001. *National Symp. on Pulses and Oilseeds for Sustainable Agriculture at TNAU, Coimbatore, 29th to 31st July.* pp.98

Mundra, S. L. and Bhati, D. S. 1991. *Indian J. Agron.* **36** (Supplement) : 294-296

Murthy, V. R., Havanagi, G. V. and Nanjappa, H. V. 1990. *Indian J. Agron.* **35** (3) : 330-331

Muthukumar, T. and Udaiyan, K. 2000 *Biol. Fertili. Soils.* **31**(2) : 114-120

Muthuswamy, P., Santhy, P. and Ramanathan, G. 1990. J. Indian Soc. Soil Sci. 38 (3) : 541-542

Myaka, F. A. 1995. *Tropical Sci.* **35** (3) : 274-279.

Naceur, M. B. and Ben Naceur, M. 1991. *Tropicultura,* **9** (2) : 77-80

Oliveira, A. P., Araujo, J. S., Alves, E. U., Noronha, M. A. S., Cassimiro, C. M. and Mendonca, F. G. 2001. *Horticultura Brasileira .* **19**(1) : 81-84

Osuji, G. E. 1990. *Soil Use and Management,* **6**(3) : 152-156

Pai, G., Bagyaraj, D. J., Ravindra, T. P. and Prasad, T. G. 1994. *Curr. Sci.* **66** : 444-445

Parasuraman, P., Mani, A. K. and Suresh, M. 2001. *National Symp. on Pulses and Oilseeds for Sustainable Agriculture at TNAU, Coimbatore, 29th to 31st July*. pp 102

Patel, M. M., Patel, A. I., Patel , I. C. and Tikka, S. B. S. 2002a *First National Symposium on Arid legumes for Food, Nutritional Security and promotion of Trade, CCS Hisar,15th to16th May*. pp. 94

Patel, M. M., Patel, A. I., Patel , I. C. and Tikka, S. B. S. 2002b *First National Symposium on Arid legumes for Food, Nutritional Security and promotion of Trade, CCS Hisar, 15th to16th May*. pp. 95

Patil, N. K. B., Kamannavar, P. Y. and Biradar, D. P. 1991. *J. Maharashtra Agrl. Universities*. **16** (1) : 111-112

Pitan, O. O. R., Odebiyi, J. A. and Adeoye, G. O. 2000 *International J. Pest Management*, **46** (3) : 205-209

Polthanee, A., Wannapat, S., Anan Polthanee, Sadudee Wannapat 2000 *Kesetsart J. Natural Sci.* **34** (2) 197-204

Pushpendra Kumar, Singh, N.P. and Kumar, P. 1990. Haryana J. Hort. Sci. 19(1-2) : 210-212

Rad, S. F., Mazaheri, D. and Banali, T. 1999. Seed and plant, **15** (2): 84-98

Raj, V. C. and Patel, R. B. 1991. *Indian J. Agron.* **36** (2) : 285-286

Rajput, A. L. 1994. *Indian J. Agron.* **39**(4) : 584-587

Ram Kishun, Ramesh Chand, Kishun, R. and Chand, R. 1989. *Indian J. Hort.* **46** (2): 268-275.

Ramaraj, B., Shannugam, N. Jalali, B.L. and Chand, H. 1990. *Proc. National conference on Mycorrhiza, Haryana Agricultural University, Hisar, 14th to 16th February* pp. 107 -108

Rattan, R.K., Neelam, S. and Balloli, S.S. 1995. *Fert. News.* **40** (4) : 31-40

Santhakumari, G., Pillai, G.R., Varghese, K. and Mathew, J. 1994. *J. Tropical. Agrculturei.* **32** (1) : 30-32

Sawant, C.V., Khanvilkar, S.A. Mahalle, S.S. and Kanade, S.V. 1992. *J. Maharashtra Agrl. Universities.* **17** (2) : 325-326

Sailaja Kumary, M. S. and Ushakumary, K. 2001 *Proc. Thirteenth Kerala Science Congress, January 2001*,Thrissur, pp 29-31

Sandhu, H.S., Puri, P.K. and Brar, S.S. 1976. *Indian J. Agron.* **21**(1): 11-16

Sangakkara, U. R., Frehner, M. and Nosberger, J. 2000. *J. Agron. and Crop Sci.* **185** (3) : 201-207

Sangakkara, U. R., Frehner, M. and Nosberger, J. 2001. *J. Agron. and Crop Sci.* **186** (2) : 73-81

Savithri, P. Poongothai, S. and Chitdeshwari, T. 2001. *National Symp. on Pulses and Oilseeds for Sustainable Agriculture at TNAU, Coimbatore, 29th to 31st July*. pp 87

Saxena, M. C., Yadav, D.S. and Singh, N. P. 1971. *Report of the Agronomic works on summer pulses at Pantnagar*, Nainital, UP.

Sekhawat, M. S. 1988. *Indian J. Agron.* **33**(3) : 341-342

Shankaralingappa, B. C., Shivaraj, B., Krishnappa. and Vishwanatha, K. P. 2000a. Karnataka J. Agric. Sci. **13** (2) : 321-325

Shankaralingappa, B. C., Shivaraj, B and Vishwanatha, K. P. 2000b. Karnataka J. Agric. Sci. **13** (2) : 295-298

Shankaralingappa, B. C., Shivaraj, B., Vishwanatha, K. P.. and Mohan, S. L. 2002 *First National Symposium on Arid legumes for Food, Nutritional Security and promotion of Trade, CCS Hisar, 15th to16tb May*. pp.97

Singh, T., Singh, S. B., Singh, B. N. and Kumar, S. 1989. *Narendra Deva J. Agric. Research.* **4** (1) : 70-73

Singh, B. P. 1991. *South Indian Hort.* **39** (6) : 381

Singh, S., Singh, D. P. and Singh, B.1991. *J. Indian Soc. Soil Sci.* **39**: 328-331

Sivaprasad, P. and Shivappashetty, K. 1980. *Agric. Res. J. Kerala* **18** : 204-207

Soddi, R. R., Sreenivasa, M. N., Chittapur, B. H. and Babalad, H. B. 1994. *J. Maharashtra Agric. University.* 19 : 459-460

Srimathi, P. Malarkodi, K. and Palanisamy, V. 1999. *Madras Agric. J.* **86** (7-9): 501-503

Srinivasan, K. and Ramasmy, M. 1992. *Indian J. Agron.* **37** (2) : 265-267

Stamford, N. P., Santosh, D. R. and Lopes, A. J. R. 1990. *Revista Bresileira de Cinencia do solo,* **14** (1) : 57-61

Syriac, E. K. and Nair, R. R. 1994. *Orissa J. Agric. Research.* **7**(3-4) : 52-56

Thyagarajan, T. R., Ames, R. N. and Ahmad, M. H. 1992. *Canadian J. Microbiology*, **38** (6) : 573-576

Thyagarajan, T. R. and Ahmad, M. H. 1993. *Biol. Fertil. Soils*, **15** : 294-296

TNAU, 2002. *Recent Advances in Irrigtion Management of Field Crops.* Centre of Advanced Studies in Agronomy, Tamil Nadu Agricultural University, Coimbatore pp. 174-175

Tomy, P.J. 1985. *Training on pulses and Oil seeds production,* Regional Agricultural Research Station, Pattambi, Kerala Agricultural University. Thrissur.

Tsigewoin Tekleab, Agarwal, I. S. and Tekleab, T. 2000. *Indian J. Animal Sci.* **17** (4) : 279-283

Vimal kumar, Bala, R., Nafees, G. and Kumar, V. 1990. *Acta Botanica Indica,* **18** (2) : 290-292

Yadav, B.D., Joon, R.K. and Rana, D.S. 2002. First National Symposium on Arid legúmes for Food, Nutritional Security and promotion of Trade, CCS Hisar, 15th to 16th May. pp.94

Yadav, D. S. 1992. *Pulse Crops.* Kalyani Publishers, New Delhi, pp. 206-210

CHAPTER – 9

COWPEA PLANT PATHOLOGY

P. P. Gupta and Vivek Gupta

ABSTRACT

This chapter presents a global perspective on fungal, bacterial, nematode and viral pathogens that directly infect cowpea foliage, stem and pods. Cowpea crop is susceptible to more than 20 viral, 30 fungal diseases, 10 bacterial and nematode diseases. Many of the fungal, bacterial and nematode pathogens are seed/soil borne in nature and primarily transmitted by seed but the most destructive diseases are viral ones and many of them are transmitted from one generation to next through seed, and thus, are generally disseminated to most cowpea-producing regions of the country. Seed borne cowpea viruses, after establishment in plantings as seed borne inoculum are typically spread within field by insect vectors (either aphids or beetle species). A brief outline is presented of the main symptoms, distribution, economic importance, epidemiology, etiology, variability, biophysical properties and mode of transmission in viruses, mode of infection and control of major diseases in India.

The major fungal diseases discussed are anthracnose, charcoal rot/ashy stem, sclerotinia rot, fusarium wilt, cercospora leaf spot, brown rust/leaf rust, powdery mildew, web blight and seedlng rot, septoria leaf spot; while the major bacterial and nematode diseases discussed are bacterial blight and root knot. The major viral diseases discussed are cowpea mosaic, cowpea yellow mosaic, cowpea mild mottle, cowpea cucumovirus, southern bean mosaic and cowpea tobamovirus.

Charcol rot (Disease of cowpea)

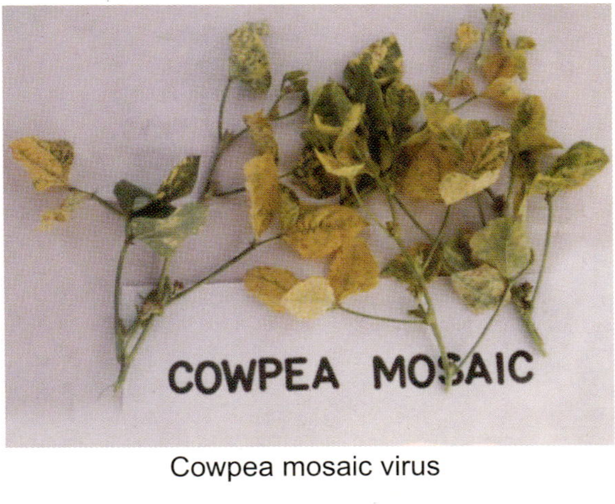

Cowpea mosaic virus

INTRODUCTION

Legumes are not only vital to the health of people and animals in a predominately vegetarian population but also play important role in agriculture by maintaining and improving soil fertility. Among legumes, cowpea [*Vigna unguiculata* L. (Walp.)] is an emerging grain legume, particularly, in the third world. Cowpea is indigenous to Africa from where it was introduced to other tropical and subtropical countries. It is now grown in India, South East Asia, West Indies, USA, Australia, parts of Europe and South America as green fodder, vegetable and grain crop. Cowpea is a major source of protein, hence, it is also known as vegetarian's meat.

In the tropical subsistence agriculture, cowpea yields are often very low for which crop loss through pests and diseases is largely responsible. The crop is susceptible to a wide range of pathogens attacking at all stages of plant growth. Some of the diseases caused by fungi, bacteria and viruses may occur in severe form during favourable weather conditions (Sharma, 1983). At present, considerable research work is being carried under grain legume programme at various research institutes and SAU's in India. Although there is substantial information available about fungal, bacterial and viral invasion of cowpea crop but in this review it has been tried to present the work done by Indian researchers during the last five decades.

IMPORTANT DISEASES OF COWPEA

1. FUNGAL DISEASES

1.1 Anthracnose (*Colletotrichum lindemuthianum*)

The disease caused by *Colletotrichum lindemuthianum* results into heavy losses in southern parts of the country (Sohi and Rawal, 1974). *Colletorichum capsici* causing anthracnose in cowpea has also been observed in Punjab (Munshi and Sokhi, 1992). Another fungus *C. truncatum* is also known to be associated with cowpea anthracnose and the disease sometimes is also called as brown blotch (Ravi and Anil Kumar, 1991b). Gupta *et al.*, (1996) reported that under artificial inoculation conditions, it can cause losses in grain yield ranging from 35 to 50 per cent. The pathogen induces leaf spot, stem blackening, pod discolouration, seed rot and seedling blight in cowpea (Rao, 1989).

1.1.1 Symptoms

The disease progresses upwards in the stem in the form of distinct lens shaped tannish pink, sunken lesions with dark red margins. The infection spreads to the leaf petiole and finally to the leaves. On the leaves, elongated lesions occur on the veins. Tannish red spots surrounded by a yellow halo develop on the leaf lamina. These spots often coalesce leading to extensive yellowing and death of the leaf. On the hypocotyls, the lesions often extend to deep cankers. Infection occurs on the peduncles and on the pods in the form of lesions similar to those found on the stem. These lesions enlarge rapidly to a diameter of one cm. Severely infected pods are mis-shapen, curled and do not attain normal size and are without seed. The affected seeds may bear brownish lesions.

1.1.2 Epidemiology and sources of infection

The primary source of infection, through seeds or plant debris of *C. lindemuthianum* has been reported (Gowda and Sullia, 1987). Wet and cool weather favours the spread of the pathogen. Damp, warm and high humidity conditions are necessary for anthracnose development. The pathogen passes the dry season in seeds and diseased plants.

1.1.3 Control

Crop rotation for 2-3 years with non-host crops may be followed. Seed treatment with carbendazim (Ramadoss and Sivaprakasam, 1989a), mancozeb (Rao, 1989), benomyl and thiram (Gaikwad and Sokhi, 1987) was effective while application of benomyl (0.2%) or mancozeb (0.2% a.i.) as spray at fortnight interval also provided good control of this disease.

The most promising means of control is the utilization of host plant resistance. Dhiman *et al.*, (1989) reported Sel.-263 as resistant cultivar against this pathogen. Fungicide resistant mutant of *C. lindemuthianum* are also reported. Naik and Anil Kumar (1992) used cultural filtrates of phylloplane fungi from cowpea of *C. lindemuthianum* resulting in strong inhibitory effect on conidial germination.

1.2 Charcoal rot/Ashy stem/Seedling blight/Root rot [*Macrophomina phaseolina (=Rhizoctonia bataticola)*]

This stem blight, caused by *Macrophomina phaseolina (=Rhizoctonia bataticola)* is most common on cowpea and other beans in India. The primary source of infection is seed and the secondary infection is through air borne conidia or sclerotia. It may attack plants at various growth stages.

1.2.1 Symptoms

The seedlings attacked by *M. phaseolina* remain stunted and some of them develop rot in the hypocotyls region and travel downwards. The roots also exhibit rotting and seedlings get dried. The fungus grow on seed coat and cotyledons and forms sclerotia. The seedling phase is also known as charcoal rot which infects seedling during emergence. In mature cowpea plants, the most common symptoms dry or wet, grayish black sunken lesions on the lower stem and the root associated with a crust like accumulate of small black sclerotia (Maholay, 1994). The lesions enlarge to form elongated cankers with grey centers and brown margins (Plate 1). Sometimes they girdle the stem, which may break longitudinally. The surface of the canker is slightly sunken and eventually split and frays. Wilting also occurs sometimes due to vascular plugging by sclerotia. The fungus commonly occurs in a complex with other fungi and nematodes. The premature senescence is associated with sclerotial stage *(Rhizoctonia bataticola)*.

1.2.2 Mode of infection

Histopathological studies of diseased seeds and component plating of artificially inoculated germinating seeds revealed that the pathogens invade the seed coat, cotyledons, plumule and radical, thus causing pre-emergence rot. Internal seed borne nature of *M. phaseolina* in cowpea had been established by Sinha and Khare (1977c). The pycnidia have been recorded under the seed coat and the intra-embryonal nature is confirmed. The disease usually results from secondary infection from air-borne conidia.

1.2.3 Etiology

On the host *sclerotia* are formed within roots, stems, leaves and fruits. They are black, smooth, hard, 100μ-1 mm in diameter (in culture 52-300 μ). Pycnidia dark brown, solitary or gregarious on leaves and stems, immersed becoming erumpent, 100-200 μ in diameter, opening by apical ostioles. Pycnidial wall multicellular with heavily pigmented thick walled cells on the outer side. Conidiophores (phialides) hyaline, short obpyriform to cylindrical, 5-13 x 4-6 μ. Conidia hyaline, ellipsoid to obovoid 14-30 x 5-10 μ in size.

1.2.4 Variability

M. phaseolina belongs to fungi imperfecti. Variability exists in such fungi on the basis of cultural, morphological and pathogenic characters. *M. phaseolina* shows variability designated as isolate M-1 and M-2 (Ratnoo *et al.*, 1997a).

Isolate M_1: Mycelial growth dark grey, aerial fluffy regular at times zonate; hyphal thin and become thicker with age, septate and profusely branched. Sclerotia formation is good, blackish, globose to round in shape and measure 49-76 μ in diameter, no pycnidial production in culture, but dark brown globose, ostiolate bodies in host tissues are developed in abundance, measuring about 141 μ in diameter.

Isolate M_2: Mycelial growth grey, aerial compact regular and fast, hyphae thin, septate, branching to form bunch like structure, sclerotia formed abundantly, dark brown, globose to round in shape and measure 45-53 μ in diameter. No pycnidial production in culture but globose ostiolate, brown bodies developed in host tissues intermingled with sclerotia and on an average measure 167 μ in diameter. Isolate M_2 proved to be more virulent compared to M_1.

1.2.5 Epidemiology

Sandhu and Singh (1998) reported that the pathogen is both internal and external seed borne in nature and can survive in the stored seeds for about 12 months. During this period, it invades the seed in cotyledons and embryo axis due to which it gets deeply seated in seeds and acts as a major source of primary inoculum in new areas. The optimum temperature required for growth of the pathogen under artificial inoculation conditions is 30°C. The young culture is more aggressive in infecting the seedlings, whereas the susceptibility of the crop also increases with the increase in age of the plant. Higher susceptibility of cowpea was associated with the impairment of its water uptake process. Ratnoo *et al.*, (1997b) suggested that higher temperature (30-35°C) and low soil moisture (40-60%) favours the disease development in cowpea.

1.2.6 Control

Crop rotation with resistant crops and mixed cropping may help reduce the disease incidence. In a study it was observed that higher doses of neem, kusam and mahuwa cakes increased the per cent germination of cowpea seed, and 1.5 per cent neem cake was most effective in reducing the disease incidence (Ratnoo and Bhatnagar 1993).

Seed treatment with Captan or Thiram (2.5 g a.i./ kg seed), Benomyl (0.1% a.i.), Bavistin (0.1%), Topsin-M or Mancozeb (0.2%) have also been proved effective (Gautam and Narain, 1996). Soil drenching with Carbendazim (0.1%) or Quintozone (0.1 %) or Thiram (0.2%) was most effective in reducing the sclerotial production, thereby, controlling the disease incidence (Ramadoss and Sivaprakasam, 1988, 1989b and 1994). Sharma *et al.*, (1999) observed that combined effect of seed treatment with Bavistin (1 g/kg seed) and foliar spray of Blitox-50 (0.03%) performed best in reducing stem blight disease of cowpea. Garg *et al.*, (1991) observed that transition metals mixed legand complexes were more effective than free legands in reducing the infection by *M. phaseolina*.

It has been well established that VAM may improve plant growth through increased uptake of phosphorus especially in soils of low fertility (Gupta *et al.*, 1999). Of five different species of Mycorrhiza inoculated in nutrient deficient sterilized soil *Glomus fasciculatum* was best for cowpea, thereby, indicating the need for inoculation of VAM fungi for deriving maximum benefit for plant growth. Gupta *et al.*, (2000) studied the interaction of *G. fasciculatum*, *Bradyrhizobium* sp. and *Azotobacter* sp. with *Rhizoctonia bataticola* and observed that inoculation of these bioagents significantly restricted the spread of the pathogen, indicating thereby that the triple inoculation of cowpea with bioagents may play important role in the management of root and stem rot of cowpea. Ushamalini *et al.*, (1997b) compared the efficacy of Bio-control agent, plant products and other soil microorganisms. The control of *M. phaseolina* was maximum when *T. viride* intensity was more in rhizosphere soil while in case of *T. harzianum* the control was maximum when present in non-rhizosphere soil. Amendment with neem cake at 150 kg/ha also registered the minimum population of *M. phaseolina* both in rhizosphere and non-rhizosphere soils. Ushamalini *et al.,* (1997c) compared the efficacy of different bioagents against *M. phaseolina in vitro* and observed that all antagonists significantly inhibited the growth of the fungus. Among them *T. viride* and *T. harzianum* were most effective. Alagarsamy and Sivaprakasam (1988) studied *in vitro* and *in vivo* the efficacy of antagonists alone or in combination of Carbendazim, Pellteting of cowpea seeds with *T. viride* either alone or in combination of carbendazim had no adverse effect on growth of both the antagonists. Seed treatement with Carbendazim and *T. viride* in combination increased seed germination and reduced post emergence mortality.

1.3 Sclerotinia rot (*Sclerotium rolfsii*)

Sclerotium rofsii is responsible for seedling decay, seedling blight, stem rot and crown rot. The wide spread nature of this pathogen has been reported by several workers throughout the tropics.

1.3.1 Symptoms

In cowpea, usually the infection begins at soil level and extends upto the hypocotyl a few centimeters at seedling stage, while stem rot and root rot occur at later stages of the crop growth (Khare *et al.,* 1977). The affected plant parts get covered with white mycelium in which mustard like light to dark brown sclerotia are embedded. Eventually the infected plant wilt out.

1.3.2 Control

The first step involves the use of healthy seeds, clean cultivation and turning of contaminated soil. Formalin and methyl bromide can be used as fumigants in seed beds. Thiabendazole @ 30-40 g a.i./100 kg seed is recommended for seed treatment. Two applications of chloroneb or Captafol @ 0.3 kg/100 l at weekly intervals may also effective.

1.4 Fusarium wilt (*Fusarium oxysporum* fsp. *tracheiphilum*)

Wilt disease is quite serious in Northern India More than six species of *Fasarium* have been found associated with cowpea seed, seedling and adult plant that cause wilt symptoms at different stages of plant growth (Sinha and Khare, 1977b). Varaprasad *et al.*, (1987) observed that concomitant inoculation of *Heterodera cajani* and *F. solani* in cowpea caused significant decrease in photosynthetic rate (43.0%).

1.4.1 Symptoms

Initially, few plants are noticed with pale yellow leaves which soon turn yellow as the plant ages. The reddish brown lesions on the tap root and subterranean stem are also observed. These later on turn dark brown and black. Gradual reduction in formation of lateral roots is also noticed. The diseased plants exhibit stunted growth, chlorosis, drooping, withering of leaves and veinal necrosis. The plant ultimately dies. The fungus invades roots and damages the vascular system. The symptoms do not appear until the plants are about six weeks old. As the plant age advances, the wilt percentage increases. The warm soil and dry condition tend to trigger irreversible wilting. The pathogen is seed and soil transmitted.

1.4.2 Control

Cation Exchange Capacity (CEC) has been found to be playing important role in management of *F. solani* in soil. The disease has been found severe in nutrient deficient soil with CEC 4.9. There is negative significant correlation between fungal population and CEC. The higher CEC nulies the effect of fungicides when used as seed treatment (Monga and Grover, 1993). Long crop rotation may be adopted, as the pathogens remain viable in soil for longer period (Sohi and Rawal, 1974). Destruction of plant debris after harvesting play important role in managing this diseases. Seed treatment with Thiram (0.3%), Captan (0.2%), singly or in combination have proved effective in controlling the disease (Hegde and Hiramath, 1987).

Pre-establishment by fungus reduces colonization by *F. oxysporum* and severity of the disease. The VAM inoculated plants possess higher level of phytoallexins. Three different phytoallexins have been identified whose concentration directly correlate with the mycorrhizal association. They are represented as I, II and III on the basis of R_f value of 0.23, 0.17 and 0.11, respectively. Among them compound III with R_f value 0.11 strongly inhibit the germination of conidia of *F. oxysporum*. This can be the mechanism of imparting tolerance against wilt disease in VAM inoculated cowpea plant (Sundaresan *et al.*, 1993). Biocontrol agents such as *Trichoderma harzianum* and *T. viride* and some plant products such as *Abutilon indicum, Delonix regia* and neem cakes also play important role in management of cowpea wilt (Usahamalini *et al.*, 1997a).

1.5 Cercospora leaf spot *(Cercospora canescens)*

Two pathogens *viz. Cerecospora canescens* and *C. cruenta* have been reported to cause leaf spots in cowpea in all the cowpea growing areas of Fiji, Brazil, Kenya, Nigeria, Zimbabwe, India, Bangladesh, Egypt, Iran, Japan, Malaysia and Thailand (Saxena *et al.*, 1998). *C. canescens* has been found more prevalent in Ibadan area of Nigeria whereas. *C. cruenta* has been found more prevalent in north central India (Chandrashekharan and Rangaswamy, 1960). The reduction in grain yield by about 20 and 40 per cent have been attributed to *C. canescens* and *C. cruenta, respectively* (Gupta *et al.*, 1998).

1.5.1 Symptoms

C. canscens produces rough circular cherry red to dark red spots upto 100 mm diameter on the upper leaf surface. The center becomes silvary to grey when spores are produced. When the spots are numerous, the leaves turn yellow and defoliation takes place. On the lower leaf surface, the lesions are of red colour. These symptoms are generally seen after the onset of flowering. The lower surface with coloured spots does not show sporulation of the fungus. *C. cruenta* spots begin as chlorosis on upper surface which becomes dotted with necrotic spots that enlarge until the whole lesion area becomes necrotic and brown coloured. Pods may also be affected. Early pod infection appears to inhibit normal development of seed, resulting in smaller and impaired seeds. In general, older senile leaves are affected more than the young, vigourously growing ones. The disease is favoured by humid weather and results into defoliation.

1.5.2 Etiology

1.5.2.1 *Cercospora canescens*

Conidiophores are in tufts, loose to dense, fasciculate, pale to dark brown in colour, measuring 3.75-50.0 x 40-207 µm in size. The conidia are obclavate to cylindrical, straight to curved with blunt tip, hyaline with 5-18 septation and 2.5-5.0 x 30.0-208 µm in size (Vasudeva, 1963).

1.5.2.2. *Cercospora cruenta*

Conidiophores are in tufts, fasciculater, erect, geniculate, subhyaline to pale, olivaceous brown, allennated, measuring 3.0-5.0 x 20.0-51.0 µm in size. The conidia are obclavato-cylindric, straight to curved, with acute tip, hyaline to subhyaline, olevaceous brown in colour with 3-9 septa and 3.0-4.0 x 28.0-103.0 µm in size.

1.5.3 Epidemiology and Pathogen Survival

The disease is favoured by humid weather. The spores are spread by wind and water splash. Pant (1989) studied the ability of its survival in plant debris and concluded that fungal conidia did not loose their germinability when the plant debris was buried at 1.5 cm depth. The natural soil environment was more conducive to conidia survival.

1.5.4 Control

Destruction of diseased debris is essential to avoid perpetuation of the pathogen. To reduce infection through seed borne inoculum, disease free seed from healthy plants should be used. Seed treatment with thiram or captan @ 2.5 g/kg seed is most effective. Both the pathogens can be completely controlled with foliar application of Benomyl (0.2% a.i.), Dithane M-45 (0.2% a.i.) or Dithane Z-78 (0.2% a.i.) at fortnightly intervals during rainy seasons (Gupta *et al.*, 1998).

1.6 Brown rust/leaf rust (*Uromyces phaseoli, U. appendiculatus*)

The disease was first reported form south India and is now one of the serious diseases of cowpea. It is caused by *Uromyces phaseoli* var. *vignae* and *U. appendiculatus* and is prevalent throughout the year in Bangalore and its surroundings but becomes severe during rainy season. Highly susceptible lines can be almost completely defoliated by mid flowering stage.

1.6.1 Symptoms

The initial symptoms appear as minute, almost white, slightly raised pustules on the leaves. In young plants, the leaves become covered with small pustules containing light brown uredospores. Plants with heavy rust infection appears to have brown tinge from a distance and wilt quickly in periods of sporadic rainfall. The diseased plants are stunted and light green in colour.

1.6.2 Epidemiology

During the dry weather, spore release is so profuse that the entire foliage is covered by a reddish brown spore dust. The disease development is favoured by high humidity, cloudy weather, heavy rains with temperature range of 22-28^{0}C and poor growth of the crop. Poor growth and gappy stand of the crop are more favourable for disease development. Under shade, the disease incidence is comparatively less.

1.6.3 Control

The early sown crop (July) shows less susceptibility compared to late sown (September) crop (Chandrashekhar *et al.*, 1989). Application of higher doses of nitrogen results in lesser incidence of disease. Moderate doses of nitrogenous fertilizers are recommended to induce dense growth of the crop. The disease can be effectively controlled by foliar sprays of 1000 ppm Oxycarboxin or Mancozeb. It is suggested that in absence of hypersensitive reaction, changes in the host plasmalamella rather than sheath formation are of primary importance in restricting further fungal growth of *U. phaseoli* var. *vignae* in immune and susceptible leaves of cowpea.

Shobha *et al.*,(1995) identified lines B-766, 774, 1484 and 1485 as highly resistant to *U. appendiculatus*.

1.7 Powdery Mildew (*Erysiphe polygoni, Sphaerotheca fuliginea*)

Powdery mildew is caused by *Erysiphe polygoni* and is wide spread in India, Bangladesh, Srilanka, Thailand, Taiwan, Africa, Australia and Venezuela. It is one of the serious diseases affecting cowpea in South India. It is also reported to be caused by *Oidium* sp. (Sohi and Rawal, 1974). In Punjab, Jhooty and Munshi (1990) reported that powdery mildew of cowpea is caused by *Sphaerotheca fuliginea.*

1.7.1 Symptoms

Small, fluffy white patches having light grayish powdery to mealy growth of fungus that appear on the upper surface of the leaves, petioles, branches, stem and pod. The affected plants are dwarfed, leaves turn yellow, curl, wither and die prematurely. The disease is favoured by cool nights followed by comparatively high temperature. It is more serious in crowded and shaded crops.

1.7.2 Control

The disease can be managed by spraying the crop with wettalbe Sulphur (0.2 %) or Afugan (0.1%) or Cosan (0.2 %) (Munshi and Sokhi, 1992; Sohi and Rawal, 1974). Excellent control of cowpea powdery mildew can also be obtained by application of Benomyl (0.2%). Application of 0.1 per cent Calixin and 0.05 per cent Kerathane as spray immediately after appearance of the disease twice at 15 days interval strikingly reduced the intensity and spread of powdery mildew.

1.8 Web blight and seedling rot (*Corticium solani, Rhizoctonia solani, Thanatephorus cucumeris*)

Cowpea web blight caused by *Corticium solani (=Rhizoctonia solani)* is wide spread in all legume growing areas of the world. Sometimes, this pathogen together with *Pythium aphanidermatum* attacks the seedlings of cowpea during the first 2-3 weeks of growth, causing seedling rot or seedling mortality. During this period, seed decay and pre and post emergence mortality take place. The pathogen also infects the leaves and young stem tissue. It can totally destroy the leaf canopy of the crop during period of heavy rains.

1.8.1 Symptoms

The initial symptoms are small, circular, brown spots on the leaves. These spots enlarge, often show concentric bandings and become surrounded by irregular shaped water soaked areas. These lesions expand and coalesce. The affected part shrivel and dries and premature defoliation can occur.

1.8.2 Mode of Penetration

Dubey *et al.,* 1989 observed that fungal penetration occurred directly under pathogen infection cushions by the development of hyphal tips in contact with the plant surface which then penetrated the cuticle. Penetration also occurred through stomatal openings.

1.8.3 Epidemiology

The pathogen has wide host range and is soil borne in nature. The humidity plays key role in the establishment and build up of this disease. The infected weeds can also act as source of infection. The pathogen can persist in the soil for many years. Walia *et al.,* (1992) studied the influence of nutrients on the pathogenic behaviour of *R. solani* on cowpea. They observed that the seedling mortality increased with the increase in application of nitrogenous and phosphorus fertilizers. Among them ammonium sulphate and calcium monophosphate had pronounced effect. The seedling rot increased with different combinations of N, P, K. Increase in inoculum from 50-200 mg/pot resulted in an increase in seedlings mortality. The pathogenicity of the isolate abated with an increase in the age of inoculum. Inoculum of *R. solani* harvested from incubation temperature of 25-30^{0}C and pH 6.5-7.5 produced maximum disease on cowpeas. Inoculum grown at 20 and 35^{0}C resulted in lower seedling mortality.

1.8.4 Control

Seed treatment with Chloroneb (2 g/Kg seed), Thiabend-azole, Tolelofos methyl or Pencycuron was found most effective in controlling the seedling rot (Kataria *et al.*, 1989; Saharan and Singh, 1994). It has been observed that the use of *Trichoderma kouingii* spore suspension twice on artificially inoculated foliage with *R. solani* can restrict the extent of disease by causing antagonistic effect and results have been found significantly better over mixture of chemical protectants i.e. Benomyl and Cacubre Sandoz. Soil amendment of micro nutrients in nutrient deficient soils also play important role in the pathogenic behaviour of *R. solani* in cowpea. The disease was reduced best by Mn amendments at 20 μg/g soil. Higher doses of Ca decreased infection and those of Mg increased it.

1.9 Septoria leaf spot (*Septoria vignae, S. vignicola*)

In India, *Septoria vignicola* was reported to cause losses in Maharashtra and Karnatka (Rao, 1963; Rawal and Sohi, 1981), whereas, *S. vignae* is reported to cause losses in Punjab (Dar and Ghani, 1993). In Karnatka region, *S. vignicola* caused serve defoliation leading to yield loss by 40 per cent (Rawal and Sohi, 1984). High humidity and temperature between 20-30^{0}C favour the development of the disease.

1.9.1 Symptoms

Initially the disease appears as small, pinhead size, reddish brown, circular, water soaked, scattered spots on both the surface of leaves which rapidly enlarge to irregular lesions having tendency to coalesce, measures 2-3 x 5-7 mm. These spots later, turn light brown with dark brown margins. Spots can be disintegrated from *Cercospora canescence* by its smaller size and more concentration. Diseased spots ultimately turn grey in the centre revealing the presence of abundant black dot like pycnidia in clusters on upper surface.

1.9.2 Etiology

Pycnidia are spherical to globose, ostiolate, 54.18-77.40 μm in diameter. Mature pycnidia produce filiform, straight and curved 2-3 septate conidia, measuring 23.22-38.70 x 1.0-1.5 μm.

1.9.3 Epidemiology

The disease is favoured by high humidity in the atmosphere and results into complete defoliation during rainy weather. It hardly takes two to three days for complete defoliation in the favourable weather. The entire foliage turns yellowish and gives a sickly appearance when the infection is severe.

1.9.4 Control

Crop rotation plays important role in management of this disease. Spryaing with Thiabendazole @225-315 g a.i./ha or Benomyl (0.2 %) or Bavistin (0.2%) or Dithane M 45 (0.2%) twice at an interval of 2-3 weeks starting from pod setting gives effective control. Use of resistant varieties like C-152, HG-171, RC-48 and V-16 are recommended (Anil Kumar *et al.*, 1994).

1.10 Seed and Seedling Diseases

1.10.1 Seed Microflora

Numerous fungal pathogens have been found associated with cowpea seeds and seedlings causing pre and post emergence mortality and blight. The fungi associated with cowpea seeds are externally or internally seed borne in nature. *Ascochyta phaseolorum, Cladisporium vignae, Corynespora cassicola, Diporthe phaseolorum, Fasarium oxysporum* fsp. *tracheiphilum* and *Macrophomina phaseolina* are major seed borne pathogens on cowpea seeds. *Aspergillus*, *Penicillium*, *Fusarium* and *Alternaria* are responsible for moulding of cowpea seeds. Healthy seeds of cowpea harboured *Macrophomina phaseolina* upto 4.99, 2.08 and 0.42 per cent in the seed coat, cotyledons and embryo, respectively compared with 13.3, 8.33 and 1.66 per cent in the same parts of discoloured seeds (Ushamalini *et al.*, 1998a). Maheswari and Mathur (1987) observed more deleterious effect of *Aspergillus nidulans* on cowpea seeds. *Fusarium oxysporum* fsp. *tracheiphilum* was detected only in seed coat of healthy seeds (2.5%) of cowpea whereas in discoloured seeds it was 11.3, 6.26 and 2.08 per cent in the seed coat, cotyledons and embryo, respectively. *Macrophomina phaseolina* caused more reduction in seed germination, root and shoot length and vigour index than *F.o.* fsp. *tracheiphilum.* At 10 per cent inoculum level, both *M. phaseolina* (85.0%) and *F.o.* fsp. *tracheiphilum* (86.3%) cause maximum infection in cowpea. Deterioration in cowpea seeds containing seed micro-organism increased with increase in storage time, temperature and RH. *M. phaseolina* caused more reduction in germination of cowpea seeds when stored at 100 per cent RH than *F.o.* fsp. *tracheiphilum, A. flavus* and *A. niger.* Germination of cowpea seeds inoculated with *A. niger* were less affected when stored at 10°C.

1.10.1.1 Effect on germination

Sharma *et al.,* (1988) examined fungi associated with cowpea and reported *Fusarium moniliforme, Penicillium* sp., *Fusarium* sp., *Collectotrichum gloeosporioides* and *Aspergillus niger* as dominant in Karnatka, while in Madhya Pradesh, *M. phaseolina* and *Fusarium equisetti* were the major seed borne pathogens causing seed rot, seedling decay, wilt and dry root rot at various stages of crop growth (Sinha and Khare , 1977b). *A. niger* and *A. flavus* adversely affected seed germination and seedling vigour when stored at room temperature. Mishra *et al.,* (1990) collected cowpea seeds from grain market and warehouses and recorded some new post harvest diseases of cowpea seeds. The organisms were *A. flavus, Chaetomium brasiliense, Rhizoctonia solani* and *Rhizopus arrhizus* causing mean losses of 8.7, 4.3, 2.4 and 30.9 per cent, respectively.

1.10.1.2 Control

Deltan treated cowpea seed can maintain viability upto 28 months (Gupta and Singh, 1990) and also reduces the number of seed microflora in storage. Seed treatment with Carbendazim (Ramadoss and Sivaprakasam, 1989c), Mancozeb (Rao, 1989), Benomyl (Gaikwad and Sokhi, 1987), Thiram (Sinha and Khare, 1977a) and Thiram, Carbendazim and Captan (Hegde and Hiramath, 1987) have resulted in better plant stand.

1.10.2 Seedling Diseases

At early stages of crop growth, a number of pathogens cause seed rot and seedling damage producing various types of symptoms resulting in mortality in certain cases. The major fungal pathogens which are responsible for causing disease at early stages of crop growth are: *Pythium aphanidermatum,*

Phytophthora vignae, *Rhizoctonia solani*, *Colletrotrichum lindemuthianum* (Khare *et al.*, 1977), *Sclerotium rolfsii*, *Fusarium oxysporum* (Sharma *et al.*, 1988), *F. solani*, *F. equisetti*, *M. phaseolina* (Sinha and Khare, 1977b), *Thelaviopsis basicola* (Anon., 1981).

1.10.2.2 Control

The management practices for seedling diseases of cowpea involve some basic precautions, like the assurance of well drained soils, removal of susceptible weeds and crop residues, maintenance of crop and plant population. Use of clean water and pathogen free soil is pre-requisite. Seed treatment with agrochemicals has always been recommended for the management of seed and soil-borne pathogens. Most common fungicides are Thiram, Captan, Benomyl or Mancozeb used in various dosages singly or in combination. Application of copper sulphate granules in water channel also regulates the seedling diseases, however, prolonged application of copper may show adverse effect on seedling growth (Khare *et al.*, 1977).

2. BACTERIAL DISEASES

Among various bacterial disease affecting cowpea, blight is most wide spread.

2.1 Bacterial blight (*Xanthomonas axonopodis* pv. *vignicola*)

The disease is widely distributed in tropical and subtropical parts of Africa, USA and India and in severe conditions, kill the entire plant. The loss in seed yield may vary from 2.66 to 92.24 per cent depending upon the nature of cultivar and the stage of infection (Kishun, 1989).

2.1.1 Symptoms

The first symptoms of this disease are water soaked dots which appear on the lower surface of the leaf. The lesions vary in size from 0.5 to 2.0 cm with the halo about 0.1-0.5 cm wide. During sever infection, complete defoliation may occur. The vascular system becomes brown. The infected stems become cankerous near the ground level and cracking occurs. The pods also show water soaked lesions which later become circular with dark green spots. The plants become stunted and bushy in appearance. Secondary infection on leaves appears as light yellow circular spots which are 4-10 mm in diameter and scattered on the lamina. The centre of these spots is necrotic and brown. The infected leaves after sometimes turn grey and fall down (Plate 2).

2.1.2 Epidemiology

The infected plant debris or seed act as primary source of inoculum, whereas secondary spread of disease is usually from the infected parts of the plant. Inoculum concentration and multiplication of pathogen play important role for symtom development in cowpea. A minimum inoculum concentration of 8.6×10^3 c.f.u./ml is necessary for symptom development when inoculated by spray and Carborundum abrasion. However, maximum disease was recorded with Carborundun abrasion at 8.6×10^8 c.f.u./ml with an incubation period of 5 days. The bacterial population declined 15 days after inoculation with necrotic tissue. The optimum RH ranges of 70-95 per cent and temperature ranges of 25-30°C (Kishun, 1989) were important.

2.1.3 Control

Sowing in the months of July is best to receive highest grain yield both under natural and artificial inoculation conditions. The disease intensity also increases with increase in plant population. Thammaiah and Khan (1995b) reported that hot water treatment at 52°C for 15 minutes was most effective in eliminating the seed borne pathogens. Treatments with mercuric chloride, antibiotics, solar heat treatments and hot water treatment for surface sterilization of infected seeds is recommended (Jindal and Thind, 1990).

In vitro studies on seed treatment with streptomycine sulphate and copper oxychloride indicated that copper oxychloride was highly effective in eliminating the pathogens from artificially inoculated seeds at all concentration under studies (Thammaiah and Khan, 1995a). Jindal and Thind (1993) observed that ratio of 49:1 of *Bacillus* sp. *E. herbicola* and *P. oxalicum* with pathogen was most effective in controlling this disease. Thammaiah and Khan (1995d) reported the presence of water soluble alkaloid in plant extracts of

Adhatoda zgylanica which was effective *in vitro* as well as *in vivo* in inhibition of growth of *X. a.* pv. *vignicola.* Root and leaf extracts treatment of infected seed for 60 minutes completely suppressed the development of disease. Root extract was more effective than leaf extract. Genotypes/ varieties such as UPC-125, V-240, 707, S-7552, Copusa-1, K-317, S-488, H-2, EC-1956, Setswana and B-342 were found resistant (Shah *et al.*, 1990; Powal and Selawe, 1993).

3. VIRAL DISEASES

The viral diseases of cowpea have recorded more attention due to their devastating effects and being a major constraint to large scale cowpea production in India. Different manifestations of mosaic are commonly observed in cowpea. Both seed borne and non-seed borne viruses, particularly, in the natural combinations cause serious losses in cowpea grain yield. Seed borne viruses have epidemiological, ecological and quarentine implications. Viruses causing diseases of cowpea include the members of alfalfa mosaic, bormo, carla, como, cucumo, hepo, poty, sobemo and tobamo virus groups (Matthews, 1982). The economically important virus diseases affecting cowpea production are described below.

3.1 Cowpea Mosaic

Cowpea mosaic disease is caused by cowpea Yellow Mosaic Virus. The disease is widely distributed in many parts of the world and yield reduction to the extent of 40-100 per cent have been reported. Two strains of CPMV i.e. CPMV-Sb and CPMV-Vu were described (Agarwal, 1964). Gilmer *et al.*, (1974) explained its seed borne nature upto 1-5 per cent but due to massive insect activity on cowpea, its incidence can build up rapidly in the crop upto 100 per cent. A large number of insect vectors such as *Ootheca mutabilis, Cerotoma variegatea, C. ruficornis, C. trifurcata Diabrotica balteata, D. undecimpunctata howardii, D. virgifera* and *Acalymma vittatum* have been reported. Beetle vectors mainly remain viruliferous for one to more than eight days.

3.1.1 Symptoms

The leaves may develop a bright yellow and light green mosaic of increasing severity in younger leaves with moderate distortion and restriction in size. Under natural conditions, two distinct types of necrotic local lesions or chlorotic local lesions are generally observed. The positive systemic reactions vary from a barley discernible green mosaic to complete death of the plant (Plate 2).

3.1.2 Mode of transmission and other hosts

The virus is readily sap transmissible, experimental host range is limited and only few hosts are known outside family leguminoseae. The diagnostic species include cowepa cvs. Black Eye and Early Ramshon and *Chenopodium quinoa.* Other hosts of this pathogen include black gram, moong bean, bengal gram, soybean, pigeon pea, winged bean and yardlong bean. Studies on the relationship of the pathogen with the aphid vector, *Myzus persicae* were undertaken. The optimum acquisition and inoculation feeding periods were 10 minutes and 24 hours, respectively. Pre-acquisition fasting of 1 hours and a minimum of 8 aphids/plant were optimum for virus transmission. In serial transmission, the virus was carried in the vector in a non-persistent manner (Nagaraju and Murthy, 1997). CPMV is also transmitted by various beetles. The chrysomelid beetle (*Ootheca mutabilis*), *Paraluperodes guaternus* (Chrysomelids) and *Namatocerns acerbus* (Curculionides) are efficient vectors. Beetle vectors remain viruliferous for 1-2 to more than 8 days. Virus infectivity was found in excrements of beetle. The transmission by two species of thrips (*Sericothrips occipitalis* and *Taeniothrips sjostedti*) and two species of grasshopper (*Cantatops spissus* and *Zonocerus variegatus*) also takes place.

3.1.3 Control

For effective disease management, the weeds that act as a alternative host should be removed and destroyed. The rouging of infected plants just after appearance of symptoms reduces the spread of the disease. The other possible method of controlling, includes vector control by use of insecticides. Spraying the crop with 0.2 per cent Endosulfan may check the insect vectors. However, application of insecticides may not achieve complete control of this disease.

The higher level of phenolic compounds and peroxide enzyme activity have been reported to contribute towards resistance to CPMV (Sangwan *et al.*, 1996). Kumar *et al.*, (1994) reported that resistance to CPMV was controlled by a single dominant gene whereas, in recent studies, Sangwan *et al.*, (1996) suggested that resistance to CPMV was governed by two dominant genes. A large number of genotypes *viz.* Vector, K-798, VITA-3, Braham, K-892, Groit, Clay K 173, Arlington, Sel 263, CS-39, 55, 82, 88, 94 TNX-294, V-16, 26, 27, 36, 154, 269, 282 and V-385 were found stable for vary degree of resistance against this disease over the locations and seasons Dhiman *et al.*, 1989. Gandhi *et al.*, 1990 observed that Blasticidin-S was the most effective inhibitor of virus multiplication (98.4%) followed by Azaguanine (70.0%), Methyl luracil (69.4%), Bromouracil (66.6%), Diazouracil (51.7%) and Nitrouracil (46.9%). Viral infectivity was reduced in callus tissues irradiated by gamma rays and X-rays by 50.8 and 43.9 per cent, respectively.

3.2 Cowpea Yellow Mosaic

The causal virus of cowpea yellow mosaic is identified as Alfalfa Mosaic Virus (AMV). The prevalence of AMV infected cowpeas is generally less than 10 per cent, although inoculation of cowpea in the field with AMV at the full bloom and prebloom stages reduces seed yield by about 15 and 50 per cent, respectively.

3.2.1 Symptoms

The naturally infected cowpea plants exhibit yellow (Calico) mosaic foliage symptoms, at times, leaf yellowing accompanied by leaflet deformation and dwarfing and plant stunting. Infected pods turn yellow without any discolouration of seeds.

4.2.2 Biophysical Properties and Other hosts

AMV is characterized by isometric and bacilliform particle containing RNA genome divided into different particles. Particles of AMV ranged 18 to 60 mm in length and width. Besides cowpea, the virus produces yellow mosaic symptoms in pigeonpea, soybean, urdbean, mungbean and systemic symptoms in chickpea, lentil, Lucerne, limabean, frenchbean, broadbean, tobacco and cucumber.

3.2.3 Control

To control the causal virus of cowpea yellow mosaic diesese, similar cultural and chemical practices may be applied as in cowpea mosaic. Till now no resistance to AMV in cowpea has been reported in India. Looking to the losses in seed yield caused by AMV, it seems essential to locate suitable sources and levels of resistance to AMV in cowpea.

3.3 Cowpea Mild Mottle

Cowpea Mild Mottle Virus (CPMMV) was first found in *Vigna unguiculata* in Ghana. Later CPMMV has been reported from other parts of world like Ivory coast, Nigeria, Kenya, Israel, India, Thailand, Indonesia and Fiji. In India CPMVV has been detected in seeds from certain cowpea genotypes (Mali *et al.*, 1986). They also characterized two groundnut isolates, one from Andhra Pradesh and other from Maharashtra.

3.3.1 Symptoms

Naturally infected cowpea plants exhibit either mild systemic mottle or no symptom but inoculated plants of susceptible cultivars local lesions on primary leaves and severe systemic chlorosis and necrosis on trifoliate leaves. Virus particles are found in mesophyll, epidermis, palisade parenchyma and in cytoplasm. Inclusions present in infected cells are unusual in shape i.e. brush like structure apparently containing virions (Mali and Thottapilly, 1986a).

3.3.2 Host range

The virus is transmissible by whitefly (*B. tabaci*) and through sap. Diagnostic experimental hosts included *Beta Vulgaris, Chenopotium murale, C. qunioa, Arachis hypogea, Glycine max, C. cajan, V. unguiculata* and *Tetrogonia expensa.* It is seed borne (2-90%) in cowpea and *P. vulgaris* (Mali *et al.*, 1989a). Properties of different isolates obtained from various workers (Suryavanshi and Mali, 1987; Mali and Nooh, 1990). Cowpea isolate is distinct from ground nut and soybean isolates and from urdbean isolate in host range, symptom types and stability of particles in sap. These isolates are found similar in serological properties and particle morphology.

3.3.3 Control

Control procedures have not been devised but selection of seed from healthy plants should prove beneficial. Seed lots suspected of containing infested seeds may not be planted near peanuts, soybean, urdbean and tomato crops. CPMMV can be managed by the manipulation of date of planting and by using field resistant cultivars. Thombre *et al.*, (1987) have reported that the disease was significantly less in early planting than in late plantings.

3.4 Cowpea cucumoviruses

Cowpea infecting cucumovirus include: cucumber mosaic virus-cowpea strain (CMV-CS), CMV bean borne strain (CMV-BS), CMV cowpea ring spot virus (CMV-RS), CMV-banding mosaic isolates (Sharma and Varma, 1975a). Cowpea ringspot virus (Phatak *et al.*, 1976), Cowpea necrosis virus (Sharma and Varma, 1975a) and a cowpea cucumovirus (Mali, 1989; Mali *et al.*, 1990). CMV CS has been reported to cause yield reduction of cowpea seeds by 14.2%. In India, CMV-BM is found more prevalent (40% incidence) than the CPNV (13% incidence). CMV-BM reduces cowpea seed yields by 40% (Sharma and Varma, 1975b). In CMV-RS the disease incidence sometimes can be seen upto 60 per cent (Patil and Mali, 1982).

3.4.1 Symptoms

Symptoms in most varieties are mild or partly masked under some conditions. Cowpea banding mosaic (CPBMV) isolate causes mosaic mottling accompanied by slight crinkling and reduction in leaf size, typical green vein banding. Diseased plants remain green for longer periods and set fewer pods under field conditions. CPNV causes mild mosaic, severe crinkling and curling symptoms. Leaf lamina of infected field plants is severely reduced giving a shiny appearance with thickened leaves and veins (Sharma and Varma, 1975a). CPRSV causes local chlorotic and necrotic lesions followed by systemic symptoms of vein clearing, chlorotic ringspots and mottle. At early stage, it shows puckering, sparse ringspot and pale yellow discolouration of veinlets. The infected plants show concentric and necrotic ringspots, veinal necrosis, necrotic streaking of stem and petiole, top necrosis, defoliation, stunting and concomitant death of the plants. Some cultivars develop systemic necrotic streaking or ringspots and mottle. The seeds are shriveled, reduced in size and show typical concentric rings on seed coat (Phatak, 1974).

3.4.2 Mode of Transmission

All the cucumoviruses are seed borne in cowpea except CPNV. However, the level of seed transmission has been found to vary among cowpea cucumoviruses. It ranges to 4-26 per cent for CMV-CS, 15-31 per cent for CMV-BM, 15-20 per cent for CPRSV and 3.5-18.5 per cent for CMV-RS (Mali *et al.*, 1987; 1989b). It has been established that the level of seed transmission varies between cucumoviruses and is dependent upon virus cowpea cultivar reactions (Mali *et al.*, 1989b). Gupta and Summanwar (1980) attributed higher percentage of seed transmission in cowpea in case of CMV-BM to the localization of virus in the plumular bud but not in the testa of seed. All the cowpea cucumoviruses except CPRSV are transmitted by aphid vectors in a non-persistent manner. Aphid vectors include *Aphis craccivora, A. euonymie, A. fabae, A. gossypii* and *Myzus persicae* (Mali *et al.*, 1990; Phatak, 1974; Sharma and Varma 1975b). *A. craccivora* is reported to be the most efficient vectors followed by *M. persicae, A. gossypii* and *A. fabae* for CMV-BM and CPNV (Chenulu *et al.*, 1968).

3.4.3 Control

The best control seems to be use of virus free seed and resistant cultivars. Sharma and Varma (1975c) reported that dry heat treatment for 15 minutes at 65°C followed by 2-4 or 8 days incubation at 30°C eliminated seed transmission of the virus without affecting seed viability. Soaking of seeds in maleic-hydrazide at 40,100 or 400 ppm for 90 minutes, 2-thiouracil at 500 or 700 ppm for 60 minutes, naphthalene acetic acid (NAA) at 40 ppm for 240 minutes and Teepol (5-10%) for 240 minutes eliminated seed transmission of the virus without affecting seed viability (Sharma and Varma, 1975c). Khalikar (1979) reports that soaking of seeds in 2-thiouracil at 500 ppm for 60 minutes, maleic hydrazide at 40 ppm for 120 minutes and NAA at 80 ppm for 20 minutes reduces CMV, seed transmission by 78, 84 and 85 per cent, respectively. Cowpea genotypes having resistance to CMV-seed transmission (CG 28, CM 11, Co-Pusa-3, K-11, L-779, 868) have been identified by Mali *et al.*, (1987).

3.5 Southern Bean Mosaic

The cowpea disease caused by southern bean mosaic sobemovirus (SBMV) was originally discovered in green house grown cowpea seedlings in Georgia. The field occurrence of this disease has been reported from USA, Ghana, Nigeria, India, Pakistan and Ivory Coast (Mali and Thottapilly, 1986b; Nain *et al.*,1994).

3.5.1 Symptoms

Cowpea seedling from infected seeds usually develop vein clearing in the primary leaves, followed by mottling in trifoliate leaves. Trifoliate leaves formed after inoculation develop vein clearing, followed by a mild to severe mottling or coarse mosaic pattern. Systematically infected leaves frequently develop a dark green vein banding along the major veins. Noticeable rugosity and a downward cupping of the leaf margins usually accompany the mottle on affected laminae. At least four responses to inoculation, including a lack of symptoms, necrotic lesions on the inoculated leaves followed by systemic necrosis, local chlorotic spots followed by systemic mosaic have been reported in different cowpea growing areas (Saxena *et al.*, 1998).

3.5.2 Host Range and Mode of transmission

In India, natural infections of sickle seha and cowpea by SBMV have been reported (Singh and Singh, 1974). However, *Cassia tora* is listed as diagnostically in susceptible plant species to SBMV (Sehgal, 1984). Occurrence of SBMV in *Vigna mungo* and SBMV-C in cowpea has also been reported from India (Mali *et al.*, 1989b; Nain *et al.*,; 1991a). SBMV-C was found to be seed-transmitted in two cowpea cultivars but not in Kuwarcheda Blackeye (KBE-1 isolate) in Maharashtra (Mali *et al.*, 1989b). Properties of KBE-1 isolate of SBMV have been determined (More, 1981). KBE-1 isolate was readily transmitted by mechanical means but not by aphid. It was also not seed-borne in cowpea and had restricted host range. It could infect five out of 20 test plant species including *Dolichos lablab*, Frenchbean (*P. vulgaris*), mungbean and urdbean. The SBMV has a very restricted host range, is transmitted through leaf beetles (Chrysomelidae) seed and inoculation of sap. SBMV is seed borne in French bean (1-5%) and more frequently in cowpea (5-40%). Evidence of embryo-transmission in French bean in equivocal.

3.5.3 Control

Control measures include use of insecticides for checking field spread of virus by beetle vector, use of SBMV free seed for planting and use of resistant cultivars. Cowpea genotypes *viz.*, RC-80, HC-2, CH-86-4 and T-9 have been reported resistant to SBMV-C in India (Nain *et al.*, 1991b).

3.6 Cowpea Tobamoviruses

Tobamoviruses causing cowpea disease are variously described from different countries. These include cowpea chlorotic spot (Sharma and Varma, 1975a), Catjang mosaic (Cappor *et al.*, 1947) and Nylon cowpea seed borne mosaic (Kulthe and Mali, 1979) from India, cowpea mosaic from the USA,; legume or cowpea strain of tobacco mosaic (TMV-CS) from Nigeria. These diseases have been reported to be caused by or related to sunhemp mosaic virus (SHMV). Cowpea chlorotic spot virus (CPCSV) incidence under field conditions has been reported from 2 to 20 per cent and yield reductions from 5 to 25 per cent from India (Sharma and Varma, 1975b).

3.6.1 Symptoms

Symptoms caused on cowpea may vary among isolates. TMV-CS causes only a mild green mottle on trifoliate leaves. CpCSV causes chlorotic spots which later coalesce on primary and trifoliate leaves, fine vein clearing in young leaves, reduction in lamina, dwarfing, early defoliation and poor pod formation (Sharma and Varma, 1975a). Catjang mosaic and other cowpea isolates cause severe mosaic accompanied by leaf malformation and reduction and dwarfing of vines in highly susceptible cowpea lines; mild mosaic or a mild chlorotic itching in less susceptible cowpea lines and a transient chlorotic itching in tolerant cowpea varieties (Capoor and Varma; 1956).

3.6.2 Host Range

Differences in host ranges have been recorded among isolates. CpCSV has a restricted host range and causes systemic mosaic symptoms in soybean, *Phaseolus lathyroides* and *Vigna vexillata;* local and systemic symptoms in *Chenopodium amaranticolor* and local lesions on guar and *Physalis floridana* but not on *Nicotiana glutinosa.* The catjang mosaic isolate causes systemic mosaic or mottle symptoms in asparagus bean, limabean, zackbean, sunhemp, soybean and local lesions in *V. vexillata, C. amaranticolor, C. murale* but not on *C. album.* Tabacco and frenchbean are not infected by CpCSV and the Catjang isolates, TMV-CS causes, systemic mosaic or mottle in Bengal bean, frenchbean, Limabean, sunhemp, Jackbean, hyacinth bean and *Centrosoma* sp. besides cowpea and local lesions on tobacco and *N. glutinosa.*

3.6.3 Biophysical Properties and Mode of Transmission

CpCSV has TIP between 80-85°C whereas, other isolates have TIP between 90-95°C : it has a DEP between 10^{-5} – 10^{-7} and LIV from 5-8 days at room temperature. No potential vector has been reported. However, a few isolates have been reported to be transmitted by various aphid species (Capoor *et al.,* 1947; Haque and Chenula, 1967). Transmission of TMV by insect vectors has also been attributed to their acting in advertently as agents of mechanical inoculation of the various tobamoviruses infecting cowpea, only three *viz.* CpCSV (4-20%), Catjang mosaic (4%) and the nylon cowpea seed borne (2.5-22.6%) have been found seed borne in cowpea. The possibility of contamination of growing plants through cultural practices is suggested (Gupta and Summanwar, 1980). CpCSV is also reported to be present in the seed plumular bud to some extent which accounts for the production of disease plants in nature.

3.6.4 Control

The best method of control cowpea tobamoviruses seems to be the use of virus free seeds for planting and resistant cultivars.

FUTURE STRATEGIES

Major pathogens of cowpea have been studied over the past 50 years in India. Although substantial information on pathogens, control measures and mechanisms of resistance have been generated from individual pathogens, however, experiments in which several pathogens were inoculated together have provided exciting new insights into pathogen-host interactions. More such studies will improve the deployment of cowpea cultivars resistant to many pathogens. Inoculation of cowpea with *Rhizobium* spp., *Trichoderma* spp. and mycorrhizal fungi may become an integral part of Integrated Pest Management practices because these organisms mitigate the effects of pathogenic fungi and nematodes. Studies on the effect of organic amendments on pathogenic survival and pesticide efficacy will contribute to the development of control packages that minimize the use of chemical pesticides. Research is needed on cropping systems and bio-control agents to explain why certain combinations of crops and micro organisms suppress plant diseases. Such complex studies should help formulate strategies for controlling most pathogens in the coming years of cowpea research.

REFERENCES

Agarwal, H.O. 1964. Identification of cowpea mosaic virus isolates. *Mededalingen van de Landbouwhoge School te Wageningen,* **64:** 53-63.

Alagarsamy, G. and Sivaprakasam, K. 1988. Effect of antagonists in combination with carbendazim against *Macrophomina phaseolina* infection in cowpea. *Journal of Biological Control,* **2** (2): 123-125.

Anil Kumar, T.B.; Chandrashekar, M.; Saiffula, M. and Veerappa, K.S. 1994. Assessment of partial resistance in cowpea to leaf spot (*Septoria vignicola* Rao). *Tropical Agriculture,* **71** (1): 36-40.

Anonymous, 1981. Pest control in tropical grain legumes, centre for overseas Pest Research, Overseas Development Administration, London, 206pp.

Cappoor, S.P. and Varma, P.M. 1956. *Indian Journal of Agriculture Sciences,***26**: 95-103.

Cappoor, S.P.; Varma, P.M. and Uppal, B.N. 1947. A mosaic disease of *Vigna catjang* Walp. *Current Science,* **16**: 151.

Chandershekhar, M.; Kumar, T.B.A.; Saifulla, M. and Yadahilly, Y. H. 1989. Effect of different dates of sowing cowpea on the severity of leaf rust caused by *Urmyces phaseoli* var *vignae*. *Tropical Agriculture,* **66** (2) : 149-152.

Chandrashekhran, S. and Rangaswami, G. 1960. *Indian Phytopathology,***13**: 96-99.

Chenulu, V.V.; Sachchidandanda, J. and Mehta, S.L. 1968. Studies on a mosaic disease of cowpea from India. *Phytopathology Z.,* **63**: 381-387.

Dar, G.M. and Ghandi, M.Y. 1993. Leaf spot disease of cowpea casused by *Septoria vignae. Indian Phytopathology,* **46** (1) : 95.

Dhiman, J.S.; Tarsem Lal and Surjan Singh. 1989. Resistance of cowpea cultivars to bacterial blight, anthracnose and cowpea mosaic disease and their effect on economic performance. *Tropical Agriculture,* **66** (1): 17-20.

Dubey, S.C.; Dwivedi, R.P. and Udit Narain. 1989. Mode of penetration by *Tranatephorus cucumeris* in some leguminous hosts. *Farm Science Journa* **4** (1-2): 138-140.

Gaikwad, D.G. and Sokhi, S.S. 1987. Detection of seed rot, root rot and seedling infection in naturally infected cowpea seeds in Senegal and their control. *Plant Disease Research,* **2**: 127-128.

Gandhi, S.K.; Bishnoi, S.S.; Jhorar, B.S. and Jatasra, D.S. 1990. Evaluation of cowpea genotypes for resistance to cowpea mosaic virus and yellow mosaic virus diseases. *Forage Research,* **16**: 70-71.

Garg, V.K,; Relan. P.S. and Girdhar, K.K. 1991. Functional effects of transition metal mixed legend complexes. *Bhartiya Krishi Anusandhan Patrika,* **6** (1): 54-58.

Gilmer, R.M.; Whitney, W.K. and Williams, R.J. 1974. Epidemiology and control of cowpea mosaic in western Nigeria. In Proceedings of the first IITA Grain improvement workshop, International Institute of Tropical Agriculture. 325 pp.

Gowda, M. and Sullia, S.B. 1987. Seed mycoflora of cowpea, field bean and soybean. *Acta Botanica Indica,* **15**: 165-167.

Gupta, A. and Singh, 1990. Viability of fungicide treated seeds of mungbean and cowpea in storage. *Seed Research,* **18** (1): 70-76.

Gupta, M.D. and Summanwar, A.S. 1980. *Seed Science and Technology,* **8** : 203-206.

Gupta, P.P.; Arora, R.N. and Gandhi, S.K. 1998. Diseases of Kharif fodder crops and their management. In: Diseases of field crops and their management (Ed. T.S. Thind). National Agricultural Technology Information Centre, Ludhiana. PP. 373-404.

Gupta, P.P; Kumar, Rajesh and Jalali, B.L. 1999. Response of different vesicular arbuscular mycorrhizal fungi on cowpea. *Plant Disease Research* **14** (1): 25-30.

Gupta, P.P; Kumar, Rakesh; Rathi, A.S. and Jalali, B.L. 2000. Effect of single and multiple inoculation of cowpea with Bradyrhizobium, Azotobacter and VA mycorrhiza on growth, nutrition and tolerance to infection by *Rhizoctonia bataticola. Indian Phytopathology.* In press.

Gupta, P.P.; Singh, J.V.; Singh, V.P. and Arora, R.N. 1996. Assessment of yield losses in cowpea against important diseases in Haryana under rainfed conditions. International conference on sustainable crop production in Fragile Environments held at CCSHAU, Hisar from Nov. 25-26, 1996. Abst. 8.6 pp 143.

Haque, S.Q. and Chenullu, V.V. 1967. Studies on the relationship of cowpea mosaic virus and its vector *Aphis craccivora* Koch. *Phytopathology Z.* **59:** 277-284.

Hegde, D.Y. and Hiramath, R.V. 1987. Seed mycoflora or cowpea and its control by fungicides. *Seed Research,* **15**: 60-65.

Jhooty, J.S. and Munshi, G.D. 1990. Mode of perpetuation of powdery mildew of pea in North-Western India. *Indian Phytopathology,* **43**: 455-456.

Jindal, K.K. and Thind, B.S. 1990. Microflora of cowpea seeds and its significance in the biological control of seedborne infection of *Xanthomonas campestris* pv. *vignicola. Seed science and Technology,* **18** (2): 393-403.

Jindal, K.K. and Thind, B.S. 1993. Biological control of bacterial blight of cowpea (*Vigna unguiculata*). *Phytopathologia Mediterranea,* **32** (3): 193-200.

Kataria, H.R.; Singh, H. and Gisis, U. 1989. Interaction of fungicide-insecticide combinations against *Rhizoctonia solani in vitro* and in soil. *Crop Protection,* **8**(6): 399-404.

Khalikar, P.V. 1979. Studies on seedborne virus diseases of Pusa Vaishaki cowpea (*Vigna unguiculata*) cultivar 95 pp. M.Sc. Thesis, M.A.U. Parbhani, India.

Khare, M.N,; Mathur, S.B. and Neergaard, P. 1977. Seed borne fungi or pulses their detection, location and importance. 18th International seed Testing Association Congress, Madrid, Spain May 6014, PP-3.

Kishun, Ram. 1989. Appraisal of losses in yield of cowpea due to *Xanthomonas çampestris* pv. *vignicola. Indian Phytopathology*, **42** (2): 241-246.

Kulthe, K.S. and Mali, V.R. 1979. *Tropical Grain Legume Bulletin*, **16**: 8-13.

Kumar, K.; Dahiya, B.S. and Rishi, N. 1994. Inheritance of resistance of cowpea yellow mosaic virus in cowpea (*Vigna unguiculata* (L.) Walp.) PP. 653-658. In Virology in the tropics (Narayan Rishi, K.L. Ahuja and B.P. Singh Eds.) Malhotra Publishing House, New Delhi, 281 PP.

Mahewhwari, R.K. and Mathur, S.K. 1987. Changes in lobia (*Vigna sinesis)* seeds due to some aspergillil. *Alexandria Journal of Agricultural Research*, **32** (2): 289-293.

Maholay, M.N. 1994. Longevity of *Macrophomina phaseolina* in different vegetables crop. *Indian Journal of Mycology and Plant Pathology*, **24**(2): 164-166.

Mali, V.R. 1989. Trends in plant virus research in western India. *Indian Phytopathologt*, **42** : 337-338 (Abstr.).

Mali, V.R. and Nooh, S. 1990. *Sixth Annual Convention IVS Dec.* 17-19, 1990. Poona University, Indian Abstr. (p 7.).

Mali, V.R. and Thottappilly, G. 1986a. National Symposium on viral Pathogensis and Immunity, V.P. Chest Institute, University, Delhi. Abstr. 47p.

Mali, V.R. and Thottappilly, G. 1986b. *Review of Tropical Plant Pathology*, **3**: 361-403.

Mali, V.R.; Markham, R.G.; Wells, B. and Pinner, M.S. 1989a. Proceedings of Fifth Annual Convocation IVS and Symposium on Epidemiology and Viral Diseases, Oct. 17-19, 1989. Abstr. Cv-II/2.

Mali, V.R.; Mundhe, G.E. and Shaikh, W.R. 1989b. Sero-diagnosis of six cowpea seed-borne viruses in India. *Indian Journal of Virology*, **5**: 44-55.

Mali, V.R.; Shaikh, W.R.; Deshmukh, R.V. and More, A.S. 1990. *International Journal of Tropical Plant Diseases*, **8**:

Mali, V.R.; Kulthe, K.S.; Patil, F.S.; Mundhe, G.E.; Dhond, V.M. and Deshmukh, R.V. 1987. Identification of source of resistance to some viruses and seed transmission in cowpea. *Indian Journal of Virology*, **3**: 99-103.

Mathews, R.E.F. 1982. Classification and nomenclature of viruses. IVth report of the International committee on Taxonomy on viruses, *Intervirology*, **17**: 1-200.

Mishra, D.; Pradhan, B.K. and Rath, G.C. 1990. Some new post harvest diseases of cowpea from India. *Orissa Journal of Agricultural Reseasch*, **3**(1): 76-77.

Monga, D. and Grover, R.K. 1993. Effect of cation exchange capacity on chemical control and population changes of *Fusarium solani* affecting cowpea. *Plant Diseases Research*, **8**(1) : 54-58.

More, A.S. 1981. Studies on a bright mosaic virus disease of cowpea. M. Sc. (Ag.) Thesis, MAU, Parbhani, 55p.

Munshi, G.D. and Sokhi, S.S. 1992. Diseases of leguminous vegetables. In Progress of Plant Pathological Research (Eds. Sokhi, S.S., Chahal, S.S. and Singh, P.P.) pp 309-320. *Indian Society of Plant Pathologists, P.A.U., Ludhiana.*

Nagaraju, N.; Murthy, K.V.K. 1997. Studies on the relationship between cowpea mosaic virus and its vector *Myzus persicae. Mysore Journal of Agricultural Sciences*, **31** (2): 170-174.

Naik, B.H.L. and Anil Kumar, T.B. 1992. Effect of cowpea leaf surface fungi on conidial germination of fungicide resistant strains of *Colletotrichum lindemuthianum. Indian Phytopathology*, **45**(3): 385-386.

Nain, P.S.; Rishi, N.; Bishnoi, S.S. and Singh, A. 1991a. International Conference of Virology in Tropics, Luckhnow from Dec. 2-6, 1991. Abstr. (1A-P-13), 23pp.

Nain, P.S.; Rishi, N.; Bishnoi, S.S. and Singh, A. 1991b. International Conference of Virology in Tropics, Luckhnow from Dec. 2-6, 1991. Abstr. (11A-P-9), 261pp.

Nain, P.S.; Narayan Rishi and Bishnoi, S.S. 1994. Growth stages of cowpea (*Vigna unguiculata*) vis-à-vis seed transmission of viruses. *Indian Journal of Virology*, **10**: 30-37.

Pant, D.C. 1989. Perpetuation of leaf spot organism of lobiya. *Indian Phytopathology*, **42**(1): 187-188.

Phatak, H.C. 1974. *Seed Science and Technology*, **2:** 3-155.

Phatak, H.C.; Diaz-Rhiz J.R. and Hull, R. 1976. Cowpea Ringspot virus: A seed transmitted cucumovirus. *Phytopathology Z.* **84** : 132-142.

Patil, F.S. and Mali, V.R., 1982. Natural occurrence of tobacco ring spot virus (TRSV) on cowpea in India. *Indian Phytopathology*, **35**: 399-403.

Powal, M.A.T. and Selawe, K.G.D. 1993. Assessment of cowpea cultivars for resistance to bacterial blight, induced by *Xanthomonas campestris* pv. *vignicola*, in Botswana. *Mededelingen van de Faculteit Landbouwweten-Schappen, universiteit Gent,* **58** (3a): 1155-1163.

Ramadoss, S. and Sivaprakasam, K. 1988. Control of root rot (*Macrophomina phaseolina*) and stemfly (*Ophiomyia phaseoli*) on cowpea.

Ramadoss, S. and Sivaprakasam, K. 1989a. Seed borne fungi of *Vigna unguiculata* and their control by seed treatments. *Madras Agricultural Journal,* **76**(1): 55-57.

Ramadoss, S. and Sivaprakasam, K. 1989b. Control of root rot of cowpea with soil application of fungicides and granular insecticides. *Madras Agricultural Journal,* **76**(2): 82-84.

Ramadoss, S. and Sivaprakasam, K. 1989c. Preserving viability of cowpea seeds with fungicides and insecticides. *Madras Agricultural Journal,* **76**(4): 235-236.

Ramadoss, S. and Sivaprakasam, K. 1994. Effect of fungicides and insecticides on the growth and sclerotial production of *Macrophomina phaseolina. Madras Agricultural Journal,* **81**(1) : 7-8.

Rao, A.N.S. 1989. Seed treatment of cowpea with mancozeb against seed and soil borne diseases. *Current Research, University of Agricultural Sciences*, Bangalore, **18** (1): 7-8.

Rao, V.G. 1963. A new species of Septoria on an economic host. *Current Science,* **32**: 367-368.

Ratnoo, R.S. Jain, K.L. and Bhatnagar, M.K. 1997a. Effect of atmospheric temperature on the development of ash-grey stem blight of cowpea. *Journal of Mycology and Plant Pathology,* **27**(1) : 90-91.

Ratnoo, R.S.; Jain, K.L. and Bhatnagar, M.K. 1997b. Variations in *Macrophomina* isolates of ash-grwy stem blight of cowpea. *Journal of Mycology and Plant Pathology,* **27**(1) : 91-92.

Ravi, K. and Anil Kumar, T.B. 1991b. Resistance development in *Colletotrichum truncatum* to carbendazim and thiophanate and growth and sporulation in resistant strains. *Indian Journal of Plant Protection,* **18**(2): 213-218.

Rawal, R.D. and Sohi, H.S. 1981. Sources of resistance in cowpea to *Septoria vignicola. Indian Phytopathology,* **34**: 504-505.

Rawal, R.D. and Sohi, H.S. 1984. Assessment of losses in pod yield of cowpea due to *Septoria vignicola*. *Indian Journal of Plant Pathology*, **2**: 59-63.

Saharan, H.S. and Singh, Sher. 1994. Influence of soil amendments on the efficacy of fungicides against *Rhizoctonia* seedling rot of cowpea. *Crop Research* (Hisar), **8**(1): 159-161.

Sandhu, Amrit and Singh, R.D. 1998. Role of seed borne inoculum in development of charcoal rot of cowpea. *Journal of Mycology and Plant Pathology*, **28**(2): 193-195.

Sangwan, R.S.; Lodhi, G.P. and Jaglan, R.S. 1996. Stable sources of resistance to yellow mosaic virus in cowpea. International conference on sustainable Crop Production in Fragile Environments held at CCS HAU, Hisar from Nov. 25-28, 1996. Abstr. 3.7. pp 49.

Saxena, Moly; Saxena, D.R.; Bhale, M.S. and Khare, M.N. 1998. Diseases of cowpea and their management. In: Diseases of field crops and their management (Eds. T.S. Thind) pp. 239-252.

Sehgal, O.P. 1984. Southern bean mosaic sobemovirus. In: Viruses of Tropical Plants (A. Brunt, K. Crabtree and A. Gibbs Eds.), C.A.B. International Wallingford, Oxoford, U.K. 489-492 pp.

Shah, Rakesh; Bhatnagar, M.K. and Mali, B.L. 1990. Reaction of cowpea germplasm to bacterial blight. *Indian Journal of Mycology and Plant Pathology*, **20**(3): 275-276.

Shama, S.; Raghunathan, A.N. and Shetty, H.S. 1988. Histopathological studies on *Rhizoctonia solani* infected seeds of *Vigna unguiculata*. *Current Science*, **57** : 945-948.

Sharma, S.R. 1983. Diseases of cowpea in India. *Indian Review of Life Science*, **3** : 105-134.

Sharma, S.R. and Varma, A. 1975a. Three sap transmissible viruses from cowpea in India. *Indian Phytopathology*, **28**: 192-198.

Sharma, S.R. and Varma, A. 1975b. Natural incidence of cowpea viruses and their effect on yield of cowpea. *Indian Phytopathology*, **28**: 330-334.

Sharma, S.R. and Varma, A. 1975c. Cure of seed transmitted cowpea banding mosaic disease. *Phytopathology Z.* **28** : 144-151.

Sharma, Bhim Sen; Bhatnagar, Kalpana and Cheema, H.S. 1999. Fungicidal management of stem blight disease of cowpea induced by *Macrophomina phaseolina*. *Journal of Mycology and Plant Pathology*, **29**(2): 276.

Shobha, Cherian; Anil Kumar, T.B. and Rawal, R.D. 1995. Evaluation of vegetable cowpeas for slow rusting resistance. *Plant Disease Research,* **10**(1): 22-27.

Singh, R. and Singh, R. 1974. *Portungallia Acta Biology,* **13** : 84-89.

Sinha, O.K. and Khare, M.N. 1977a. Chemical control of *Macrophomina phaseolina* and *Fusarium equisettii* associated with cowpea seeds. *Indian Phytopathology,* **30**: 337-340.

Sinha, O.K. and Khare, M.N. 1977b. Seed borne fungi of cowpea and their significance. *Indian Phytopathology,* **30**: 469-472.

Sinha, O.K. and Khare, M.N. 1977c. Site of infection and further development of *Macrophomina phaseolina* and *Fusarium equisetti* in naturally infected cowpea seeds. *Seed Science and Technology,* **5**: 721-725.

Sohi, H.S. and Rawal, R.D. 1974. Control of cowpea diseases. *Indian Horticulture,* **19**(1): 15-19.

Sundaresan, P.; Raja, N.V. and Guhasekaran, P. 1993. Induction and accumulation of phytoalexins in cowpea roots infected with a mycorrhizal fungus *Glomus fasciculatum* and their resistance to *Fusarium* wilt disease. *Journal of Biosciences,* **18**(2): 291-301.

Suryavanshi, A.P. and Mali, V.R. 1987. 3rd Annual Convocation IVS. Calcutta, Dec. 11-13, 1987, Chittaranjan National Cancer research Centre, Calcutta Abstr. FR-40.

Thammaiah, N. and Khan, A.N.A 1995a. Effect of seed treatment on *Xanthomonas campestris* pv. *vignicola* causal agent of bacterial blight of cowpea. *Advances in Agricultural Research in India,* **4**: 93-102.

Thammaiah, N. and Khan, A.N.A. 1995b. Studies on control of *Xanthomonas campestris* pv. *campestris* in cowpea seeds by hot water treatment. *Advances in Agricultural Research in India,* **3**: 194-198.

Thammaiah, N. and Khan A.N.A. 1995d. Effect of extract from *Adhatoda zcylanca* on *Xanthomonas campestris* pv. *vignicola* causing bacterial blight of cowpea. *Advances in Agricultural Research in India,* **4**: 109-117.

Thombre, P.A.; Mali, V.R. and Navgire, K.D. 1987. Annual Review. Report of AICRP on soybean cash crop. Science sub-committee, AGRESCO April, 17-18, 1987, MAU, Parbhani, India. 78-82.pp.

Ushamalini, C.; Rajappan, K. and Gangadharan, K. 1997a. Control of cowpea wilt by non-chemical means. *Plant Disease Research,* **12**(2): 122-129.

Ushamalini, C.; Rajappan, K. and Gangadharan, K. 1997b. Management of charcoal rot of cowpea using biocontrol agents and plant products. *Indian Phytopathology*, **50**(4): 504-507.

Ushamalini, C.; Rajappan, K. and Gangadharan, K. 1997c. Inhibition of *Macrophomina phaseolina* and *Fusarium oxysporum* f.sp. *tracheiphilum* by antagonists under *in vitro* conditions. *Plant Disease Research*, **12**(2): 168-170.

Ushamalini, C.; Rajappan, K. and Gangadharan, K. 1998a. Effect of seed borne fungi on germination, root and shoot length and vigour index of cowpea. *Plant Disease Research*, **13**(2): 109-114.

Varaprasad, K.S.; Raghuveer Polisetty and Sethi, C.L. 1987. Effect of separate and concomitant inoculation of *Heterodera cajani* Koshy, 1967 and *Fusarium solani* on cowpea growth and photosynthetic rate. *Tropical Grain Legume Bulletin*, **34**: 29-31.

Vasudeva, R.S. 1963. Indian Cercosporae, ICAR, New Delhi, India, 245pp.

Walia, G.S.; Sunder, S. and Grover, R.K. 1992. Influence of nutrients on pathogenic behaviour of Rhizoctonia solani on cowpea. Indian Journal of Mycology and Plant Pathology, 22(2): 170-177.

CHAPTER – 10

COWPEA ENTOMOLOGY

Satya Vir

ABSTRACT

This chapter deals with the insect pest complex of cowpea, their seasonal incidence, extent of damage caused by them, screening for resistance to pest attack and their management in arid and semi-arid region of Rajasthan. Field trials to study the extent of losses due to pest complex have revealed a yield loss of 15.16% to 71.91% in different cultivars of cowpea. Early sowing preferably in the 1st or 2nd week of July proved the best, where the incidence of jassids, whitefly and weevils was minimum. The insect population increased with delay in sowing which not only affected the normal growth of plant but also decreased the yield. The study on distribution, symptoms and damage, their biology and control of individual insect viz. whitefly, jassids, leaf damaging weevil, pod bores, cutworms, hairy caterpillars, whitegrubs, termites, grasshoppers and bruchids have been recorded in detail. The pest problems in the storage of cowpea seeds, extent of damage under improper storage conditions and their management has also been reported. The write-up also deals with the effective, economical and ecofriendly method for management of insect pest complex of cowpea.

INTRODUCTION

There is a great deal of importance attached to cowpea crop, as the same is used both for fodder and grain purposes.

The productivity of the crop is low. One of the primary causes of low yields, is the damage done by the insect pests (Booker, 1965a). Almost every part of cowpea plant is infested by one insect species or another. Booker (1965b) has listed as many as 85 species, which cause injury to cowpea. However, only a few of them are considered as the major pests of cowpea. A feature of the pest complex of grain legumes is the apparent lack of host specificity of some of the important economic species (Taylor, 1964). Besides, many insects of arid and semi-arid regions, hairy caterpillar (*Diachrysia oblique* Walk) is known to infest cowpea from seedling to flowering stage while blue butterfly (*Euchrysops cnejus* Fab.) and pod borer (*Maruca testulalis* Geyer) have been reported to damage flowers and pods of cowpea crop. The seasonal fluctuations in population, extent of damage, symptoms and biology of major pests, screening of cultivars to pest attack both in field and storage and their control mechanisms under arid/rainfed conditions is summarized in this review.

Seasonal population fluctuation of insect pests in cowpea

In Rajasthan, cowpea is exclusively grown in the kharif season from July to October. A diagrammatic representation of major insects pests attacking cowpea crop under arid and semi-arid condition is given in Fig. 1.

Population of hairy caterpillar sporadic in appearance is maximum during the second and third week of July and their second attack depends upon the continuous rainfall during July (Satya Vir and Singh, 1985a). Jassids appear on the crop during second and third week of August and their population generally roses to the peak within a fortnight. The member of jassids during peak period may vary from 8.3 to 17.50 per catch. The whitefly incidence is selectively negligible on cow pea as compared to moth bean and moong bean crops. However, this insect, also acts as a vector for yellow mosaic virus (YMV). Chlorophyll pigments of the leaves are destroyed due to YMV infestation, the photosynthesis is severely hampered ultimately resulting in low productivity of the crop. Weevils attack the crop in the middle of August but disappear in 15-21 days by first or second week of September. Termite attack is more severe, when there is dry spell of at least 25-30 days; while grasshopper are found as minor pests during the entire cropping season. Aphid infestation depends on the climatic conditions such as high humidity, cloudy weather and low sunshine hours during August to September. Heavy attack of aphids may result in 22 to 62% loss in yield depending upon the crop stage and the intensity of aphid attack (Satya Vir and Singh, 1985a).

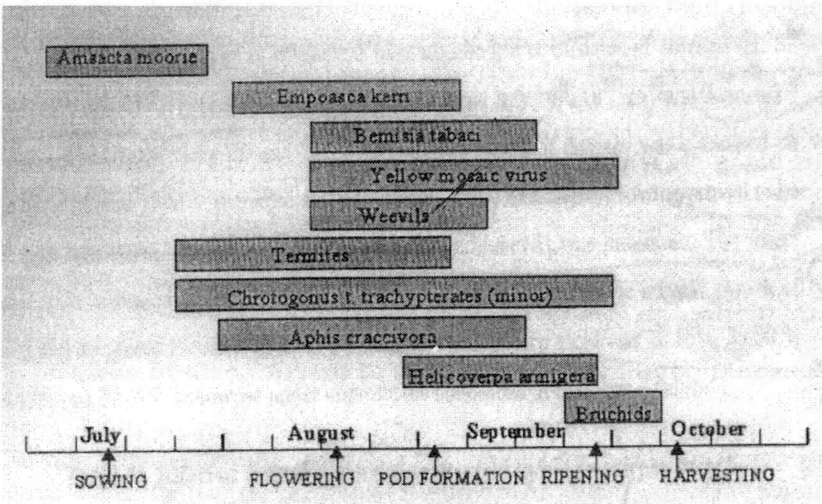

Fig. 1. Periods of attack of pest complex on cowpea crop during kharif season in arid & semi-regions of Rajasthan

Effect of sowing dates on the incidence and extent of damage

Effect of varied sowing dates viz., 8, 16, 23 and 30 July in the first year and 10, 18, 25 July and 2 August in second year on the extent of damage on the of seed yield in protected and unprotected plots revealed that first sowing dates proved the best, recording minimum incidence of jassids, whitefly and weevils (Satya Vir and Singh, 1985a). The insect population increased with delay in each sowing, which affected the growth of plant and decreased the seed yield. The seed yield decreased by 22.15, 39.30, 40.62 and 62.10% under unprotected conditions during first year (Table 1).

Table 1. Effect of sowing dates on grain yield of cow pea.

Sowing dates	Grain yield (q ha^{-1})		Yield loss (%)
	Protected	Unprotected	
	19		
July 8	9.27	7.21	22.15
July 16	8.52	5.17	39.30
July 23	6.37	3.78	40.62
July 30	2.48	0.94	62.10
	19		
July 10	11.43	1.40	87.70
July 18	9.46	4.80	49.20
July 25	7.95	4.46	43.90
August 2	1.49	0.89	46.50

The results revealed that yield losses due to infestation varied from 22.15 to 62.10% during the first and 43.90 to 87.70% during the second year. In general, late planted crop suffered heavy losses due to insect attack than is comparison to the crop sown in the first or second week of July. Climatic conditions such as humidity, sunshine hours and cloudy weather during August may be the important factors influencing early appearance and multiplication of aphid population in cowpea crop.

Screening of cowpea varieties for their susceptibility to pest complex

Thirteen promising varieties of cowpea screened for their susceptibility to pest attack revealed that the yield loss varied from 26.44 to 69.84% (Satya Vir and Jindal, 1982). The cv. J.C. 5 with 26.4% damage was least susceptible to the pest, whereas the maximum yield loss was observed in the cv. viz., C.M. 11, K-39 and V-37 (66.22 to 69.84%). Popular varieties grown in this region i.e. J.C. 10, V-7, C-152 were moderately susceptible to the insect infestation (Table 2). Evaluation of 200 varieties of cowpea for resistance to *Aphis craccivora* carried out by Messina *et al.*, (1985) revealed that three varieties resistant in Western Africa were observed highly susceptible to a population of aphid from southern USA. Jayappa and Lingappa (1988) also screened 408 entries of cowpea for resistant to *Aphis craccivora* and isolated MS-370, TVU-2740, P-912 and P-1475 consistently resistant to aphid attack. Ahmed *et al.*, (1990) screened 46 genotypes for tolerance to *Aphis craccivora* and reported considerable variation in seed colour and days to 50% flowering.

Table 2. Grain yield of cowpea under protected and unprotected conditions

Variety	Grain Yield (kg ha^{-1})		Yield loss
	Protected	Unprotected	(%)
J.C. 5	459	363	26.44
Copusa-1	287	165	47.17
H.G. 22	359	265	35.47
V-37	321	189	69.84
C.G. 69	384	247	55.46
V-7	506	367	37.87
C-152	496	355	39.71
C-03	436	295	47.79
V-16	469	323	45.20
J.C. 10	542	394	37.56
R.S. 9	522	349	49.57
C.M. 11	497	299	66.22
K-39	550	328	67.88

C.D. at 5% (Var.), 138.180; C.D. at 5% (Treat) = 44.209

Cowpea cultivars against jassids and pod borers

Observations on thirty-four entries of cowpea were recorded for jassids, pod borer infestation and percentage loss in seed yield were recorded under protected and unprotected conditions. The cv J.C. 5 and J.C. 10 had minimum number of jassids (below 0.40/leaf), followed by HFG 42-1, FS-68, CO 2 (0.56 to 0.70/leaf) while Copusa-1 and V-7 had maximum number of jassids (1.2/leaf). Results on pod borer infestation revealed that varieties Copusa-1, H.G. 22, K-39, CG-28, PTB1, 25/8/1 25/13-1-10-78 had maximum pod borer infestation (6.12 to 9.09/leaf); and J.C. 5, J.C. 10, CG-28-488-9-1-1, 25/3/21, Sel-1476, CO_3 and CO_2 escaped from pod borer infestation. Per cent loss in yield due to insect pest was minimum (less than 5%) in J.C. 5, J.C. 10, HFG 42-1 and FS-68 while it had maximum in V-7, HG-22, PTB-1 and Copusa-1

(32.8 to 37.1 %). Based on three years results J.C. 5, J.C. 10, HFG 42-1 and FS-68 were considered least susceptible.

Distribution, symptoms and biology

Foliage pests

Whiteflies: Bemisia tabaci Genn. (Hemiptera: Aleyrodidae)

Distribution: This insect is common in pulse growing states of Rajasthan, Gujarat, Madya Pradesh, Haryana and Andra Pradesh

Symptoms & damage: This insect is of considerable importance because not only does it feed on the plants, but also transmits the yellow mosaic virus (YMV). The first symptom of the disease appears on the young spectra or spots. In the reverse cases the entire leaf becomes yellow and the yellow patches become brownish exhibiting necrotic areas. The transmission period of the disease is very short and the leaves get infested with YMV even if the insect comes in the contact with the leaf surface for a minute (Satya Vir, 1985). Since the chlorophyll pigment of the leaves is destroyed, the photo-synthesis is greatly affected and resulting in low productivity of the crop. The yield losses of 12.62 to 65.21% in different moth bean cultivars due to whitefly and YMV have been reported by Satya Vir (1983a and 1985).

Biology: This tiny, pale and soft bodied fly measures only 1 mm long. Eggs are laid singly on the under surface of the leaflets. The freshly emerged nymphs which are sluggish, stay on the same leaf till they are fully grown up. The nymphs moults thrice and their nymphal and pupal periods are about 2 weeks during summer but it is significantly increased in winters. Flies multiply rapidly during warm and humid conditions (Satya Vir, 1985).

Control: Field experiments on efficacy of insecticides in controlling whilefly and checking YMV revealed that monocrotophos, fenitrothion, methyl-o-demeton, dimethoate, permethrin, disulfoton, phorate and carbofuran significantly reduced the whitefly population and were found effective in reducing the incidence of yellow mosaic disease. (Satya Vir 1983c, 1985, 1998). Monocrotophos @ 0.25 kg a.i./ha applied 3 times at 15 days interval starting at 15th day of sowing was effective than other insecticides and has been recommended for the control of whitefly and yellow mosaic disease on moth bean crop. Application of disulfoton, phorate and aldicarb at the rate of 1.0 to 1.25 kg/ha at the sowing time keeps the whitefly population in check up to 4 weeks of plant growth. Pai and Dhuri 1991 observed 0.04% dimethoate as most effective for control of *Bemisia tabaci* in cowpea.

JASSIDS: *Empoasca kerri* Pruthi (Hemiptera:Cicadellidae)

Distribution: This is serious pest and causes extensive damage to almost all the kharif pulses. In India, *Empoasca kerri* is the most common species on legumes. There are some wild species also that act as alternate hosts to jassid.

Symptoms: It sucks the leaf sap leaflets become cup shaped and yellow at the edges. With heavy infestation, the leaves turn brown, curl from the edges and dry. This type of damage is referred to as 'hooperburn'.

Biology: A number of eggs are laid singly on leaf tissue on the lowerside of leaflet hatching out in 5-7 days. The nymphs are found on both upper and lower leaf surfaces but generally avoid direct sunlight. The nymphs resemble adult, but have no wings and run side ways when disturbed. One generation is completed in 2 weeks under optimum conditions.

Control: Field experiments conducted to assess the comparative efficacy of twelve insecticides for the control of green jassids *E. kerri* revealed that Methyl-o-demeton, fenitrothion (0.03%) phosphamidon, monocrotophos and permethrin (0.15%) were the most effective. Momocrotophos, methyl-o-demeton quinalphos, carbaryl and dimethoate also provided good protection up to 21 days (Satya Vir, 1983b). The use of disul-foton or aldicarb as basal application at the time of sowing, followed by a monocrotophos spray on four or five weeks old crop (Gupta, 1970) or with carbofuran as a seed treatment (Saxena *et al.* 1975) has also been recommended for the control of jassids. Application of aldicarb as a seed coating at 1.0 kg a.i./ha has also been found highly effective in reducing the incidence of *E. kerri* and, therefore, increasing yield of green fodder seeds (Ram *et al.*, 1986).

APHIDS: *Aphis craccivora Koch* (Hemiptera: Aphididae)

Distribution: Aphids is a sporadic but serious pest of kharif pulses in India. Several species of aphids are recorded on cow pea but *Aphis craccivora* Koch is most common in Rajasthan and Gujarat. Aphids in kharif seldom thrive long, probably due to high temperature and the activities of natural enemies.

Symptoms: The colonies of aphids are seen on young stem, leaflets and pods. Young seedlings become twisted and heavy infestation can cause wilting. The sap sucking activities severely check the plant growth. Damage is caused by both nymphs and adults. The loss of 43.9 to 87.70% in yield due to aphid attack has been recorded under favourable conditions (Satya Vir and Singh, 1985a).

The adults of *Aphis craccivora* are black and shiny, measure up to 2 mm long and some are winged. The nymphs are smaller with light wax covering that makes the look grey and dull. These aphids produce a new generation in a week and the infestation build up is very quick.

Control: Aphids can be effectively controlled by spraying 0.03% dimethoate, 0.07% endosulphan, 0.04% monocrotophos or 0.1% lindane at @ 700 L/ha (Satya Vir 1983e and Saxena, 1978). Natural enemies such as lady birds (*Coccinella septempunctata)* and syrphid larvae, *(Ischieden scutellaris)* have been found effective in controlling the population of *Aphis craccivora* in summer (Saxena *et al.*, 1970). Pai and Dhuri (1991) have reported 0.04% dimethoate as most effective in controlling *A. craccivora* and *Bemisia tabaci* in cowpea.

LEAF DAMAGING WEEVILS

Myllocerus undicimpuntulatus maculates Desb. (Coleoptera: Curionidae), *Myllocerus lactiverus* Marshall, *Myllocerus cardoni* Mscl, *Myllocerus uniformia* Marsh, *Myllocerus discolor* Boli

Distribution: Leaf damaging weevils are found throughout India. They feed on kharif pulses, cereals like bajra and it is also a common pest of cotton.

Symptoms: These insects chew the leaflets, generally at the margins in case of *Cyrtozemia* spp. and make central holes in case of *Myllocerus* spp. The grubs feed underground on the root seedlings where one grub can destroy 9 seedlings in 40 days. The adults feed on leaves, buds, flowers and young bolls.

Biology: The pest becomes active from early kharif season and passes winter in the adult stage, hidden in debris. The weevils appear in April-May and lay ovoid light yellow eggs in the soil. The female lays on an average 360 eggs over a period of 24 days which hatch in 3-5 days. The young grubs feed on the roots. The grubs complete their development in 1-2 months. They pupate in the soil inside earthen sells forming creamy white pupae which change into adults a week, the adults live for about 8-20 days.

Control: The adults and larve do not cause sufficient damage to merit separate control measures. But under favourable environmental conditions there is high density of pests per plant. The use of malathion dust @ 20-25 kg/ha is recommended (Satya Vir, 1983e).

POD BORER: *Helicoverpa armigera* Hubner (Lepidoptera: Noctuidae)

Distribution: *Helicoverpa armigera* is cosmopolitan and widely distributed in India. It is serious pest of all kharif pulses.

Symptoms: The larvae are polyphagus but they prefer to feed on the pods of gram, red gram, green gram and cow pea. They feed on the foliage when young and on the seeds in the later stages and thus, reduce the yield. A single larvae may destroy 30-40 pods before it reaches maturity. The pest also destroys buds, flowers and pods. If flowers and pods are not available, they feed upon leaflets; leaving the veins. Usually developing and partly matured seeds are eaten completely.

Biology: The insect lays their small white eggs, usually singly, on the upper and outer surfaces of leaves, flowers, pods and stems. The young larvae feed by scraping green tissue and the older larvae chew voraciously into buds, flowers and pods leaving characteristic round holes. The large larvae are yellow, green, pink, orange, brown or black, but all have characteristic light and dark stripes along each side of the body. The pupation takes place normally in the soil or in plant debris. One generation is completed in 4-5 weeks under favourable conditions.

Control: The insect can be effectively controlled by spraying with 0.07% endosulphan @ 700 1/ha (Saxena, 1978). Phosphemidan (0.04%), carbaryl, (0.1%), endosulfan (0.05%) are equally effective for the control of pod borers on cowpea crops (Jaiswal and Patel, 1993).

CUTWORM: *Agrotis ipsilon* Hufnagel (Lepidoptera: Noctuidae)

Distribution: This insect is wide spread, polyphagus and reported all over India.

Symptoms: Seedlings are cut through at or below ground level.

Biology: The large moth (25 mm long) have brown forewings and pearly hind wings with a brown margin. The crown coloured eggs are laid singly on the plants or soil surface. The larvae feed on leaves, are grey black and measure around 40-45 mm in length. They hide beneath the soil surface during daytime and feed at night. Pupation takes place in the soil. One generation is completed in 5-6 weeks.

Control: The pest can be controlled by mixing chlorpyrphos at the time of sowing or dusting methyl-perathion dust @ 20-25 kg/ha on crop.

RED HAIRY CATERPILLARS: *Amsacta moori* Butter (Lepidoptera: Arctidae) *A. albistriga F.*

Distribution: The red hairy caterpillar is a sporadic pest of kharif crops in arid and semi-arid regions of Rajasthan. It is a ployphagus pest and has been observed feeding on several legumes, cereals, certain rainy season vegetables, ornamental plants and weeds in arid zone of Rajasthan. *Aerva pseudotomentosa, A. persica* and *Heliotropium sublatum* were observed to be favourite host plants of this insect (Satya Vir and Singh, 1985b).

Biology: The moth generally emerges after the first shower of monsoon and are attracted towards light. Females are slightly bigger than the males and both have pink dots on wings. The eggs are laid on the lower surface of leaves preferably on the weeds. A female normally lay about 160-265 eggs. The larval period extends from 25-35 days whereas, the average length of the adult period varies 4 to 7 days.

Control: Quinalphos (0.075%), endosulphan (0.1%) and dichlorovos (0.025%) were found significantly superior against 3rd and 5th instar larvae of hairy caterpillars. However, the percentage larval mortality in 5th instar was less compared to 3rd instar larvae. Monocrotophos (0.025%) and fenitrothion (0.025%) were found equally effective after 48 hrs and 72 hrs of treatement.

WHITEGRUBS

Distribution: Both whitegrubs and termites comprise the substerranean group of dominant insect pest fauna. Considering the race for the survival, soil may be interpreted as an ecological niche where species have escaped over-crowding existing above the surface. Population of these subterranean species depend on qualitative stratification of the soil. The whitegrubs or cockchafer grub prefer soils rich in decaying organic matter. Loose soils, in general, are preferred for subterranean movements by cockchafers while the clay is preferable for gallery making by the termites.

Nature of Damage: The grubs are active subterranean root feeders and positive gertropic while the adult cockchafer beetles are nocturnal foliage nibblers. In endemic tracts, the damage is regular, and the problem starts particularly where organic matter is applied before it is well rotted. The infestation may commence May and continue till October in plains. The young grubs soon after hatching start damaging the root region at any stage of the plant. Generally the roots show abrupt cuts in tender plants. Consequently, the young plants appear wilted or drying up and easily pulled off, the older ones initially turn yellow and ultimately their foliage wilers. Most of kharif crops are susceptible. Generally, the infestation may increase with age of the crop. The pest has been specifically serious to leguminous plants because of the tap root system with fine rootlets and nodules compared to the plants of cereals, having adventitious root system. A single grub is capable of destroying a number of plants. Up to 6 - 7 grubs have been reported around the plant with soil depth of 7-10 cm. During pod formation, the shell as well as the kernels are damaged.

Control: For managing these insects, killing of adults is an effective way. The beetles are attracted toward light in night which may be trapped and destroyed by burying, burning or dipping in insecticidal or oil emulsion solution. However, the traps should be employed at a large scale co-operative basis, otherwise more harm will be caused than profit, if isolated efforts are made. Dusting the host plants with insecticide (methyl parathion 2%) or spraying (methyl parathion/ quinalphos/denithothion 0.05%) can also kill beetles feeding on these plants.

Soil ploughing exposes the grubs to natural enemies like crows and other birds. Additional cross or deep ploughing will expose more of the grubs to birds. Soil treatment with insecticides is a costly practice but may have to be resorted to at times when none of the above cited measures were applied. The grubs are quite big in size and, therefore, normal dose of insecticides has to be increased many fold for these insects. Phorate 10% granules has proved quite effective but is quite costly considering its requirement of 25 kg/ha.

TERMITES

Distribution: Semi-arid regions suffer the maximum damage of termites and grasses in particulary are attacked the most because of high cellulose content in them. Although the losses in irrigated fodder crops are far lesser compared to the totally rainfed or irrigated ones but if the time interval between the irrigation is long, termite attack may increase.

Control: For effective control of termite, preparation should be made in anticipation before establishing forage crops in the field. Mixing of soil insecticides like Chlorpyriphos at the rate of 25 kg/ha of dust formulation, before or at the time of sowing/transplanting renders the treated field safe from termite attack. In case of isolated patches of termite injury are seen, emulifiable concentrate of Chlorpyriphos may be applied either with irrigation water or mixed with sand. In the latter case, irrigation should invariably follow insecticidal application. Under no circumstances, any of the insecticides cited above be used for termite control by spraying or dusted on forage crops.

GRASSHOPPERS

Distribution: One of the most common insect seen almost around the year or farage crops is this group of insects. More than twenty species of grasshoppers are found in the arid and semi-arid regions of Western Rajasthan. However, grasshoppers are not always considered as the major pest. In fact it's only at certain times that these insects inflict economic injury to the crops. Under favourable climatic conditions in some years the grasshopper population may shoot up. Normally, the population of hoppers does not exceed a certain level barring a few species. Natural enemies of grasshoppers like lizards, birds etc. keep their population checked.

Control: For protecting the crops from the grasshoppers, soil work may be done. The eggs laid by them may be exposed and destroyed. Dusting of insecticides may be done in the younger stage of crop but near cutting stage it should be avoided.

BRUCHIDS: *Callosobruchus maculatus* Fab., *C. Chinensis* L., *C. Analis* F. (Coleoptera: Bruchidae)

Distribution: The bruchid, *Callosobruchus maculatus* Fab. is common on all the pulses in tropics and substropics. They damage both in the field and stored seeds. The other species commonly found in stored seeds in India are *C. chinesis* and *C. analis*.

Symptoms and Losses: The pest attack nearly mature and dried pods. The round exist hole and the white eggs on the pod walls are conspicuous. Infested stored seeds can be recognized by the eggs on the seed surface and the round exhibit holes with the 'flap' of seed coat. Actual figures on losses to the growing crop are sparse. Prevett (1961) recorded that pest complex affected 3.1 to 11% of pods collected as they are dried. Phelps and Oosthuzen (1958) recorded only 1.9% damage of freshly picked cowpea pods *by C. chinensis*. However, such levels of pod infestation by potential storage pests represented a serious threat.

Biological studies of Bruchids

Oviposition response and development of *C. maculatus* on different host species

Oviposition response and development of *C. maculatus* studied on eight seed materials (Satya Vir and Jindal, 1981) revealed that arhar (80.66 eggs) was most preferred for oviposition followed by dolichos (54.66), cowpea (50), blackgram (46.0), moth bean (33.33), moong bean (22.33) rice bean (22.0) and guar (16.66) The grubs failed to penetrate the seed coat of guar and rice bean and there was no development on these seed materials. The development period was minimum on moong bean and moth bean and maximum in case of dilichos. Emergence of beetles was maximum on cow pea followed by moong bean, moth bean, black gram, arhar and dolichos. The study concluded that moong was the most preferred host followed by moth bean, arhar, blackgram, cowpea and dolichos.

Oviposition response and development of *C. maculatus* on different cow pea seeds

The study revealed that beetle was found guided in its oviposition on preference by colour, texture and volume of seed grain (Satya Vir, 1980). Bright coloured smooth surfaces with greater seed volume were much preferred for oviposition. Egg laying was minimum on wrinkled surfaces. Female tended to avoid adding second egg as long as there were unused seeds. The development and percentage emergence of adults was not affected on the grain preferred for oviposition. The study on the comparison of data on nutritional value of food and seed consumed, revealed that the amount of food consumed per grub increased with the increase in protein value of seed. The protein value of different varieties of cowpea varied from 21 to 31.5 per cent. The comparative resistant varieties contained only 21 to 24.5 per cent protein compared to most susceptible ones containing 28 to 31.5 per cent protein. The study thus, concludes that it is not the morphological characters of the seeds but the nutritional value of seeds that governs the mechanism of resistance in cowpea to the attack of *C. maculatus*.

Multiplication of *C. maculatus* and loss in weight

Multiplication of *C. maculatus* on mothbean and cowpea seeds, percentage infestation of seed and loss in seed weight was studied at four levels of infestation for a period of six months (Satya Vir, 1989). Population on both the seeds increased sharply after one month of storage. Increase in population on moth bean seed was 510, 768, 996 and 1258 with the infestation of 21, 40, 68, and 81 per cent seeds and a loss of 24, 39, 49, and 58 per cent seed weight while in cowpea it was 417, 805, 912 and 1027 insects with the infestation of 19, 21, 30, 38 per cent weight after six months of storage initially infested with one, two, three and five pair of insects, respectively.

Screening of cowpea against pulse beetle

Seventy-seven cultivars of cowpea were screened for their relative susceptibility to the attack of *C. maculates* (Satya Vir, 1981, 1982a, 1982b, 1983d, 1984). The results revealed no significant difference among the cultivars for average number of eggs laid and the percentage infestation of seeds. Cultivars reaction in regard to food consumed per grub, percentage loss in seed weight and the emergence of beetles was significant. All the cultivars were compared on the basis of average food consumed per grub and the percentage loss in seed weight. The cvs viz., J.C. 5, Sel-1476, V-38, 25-8-2-2, V-3, C-152, 647-1, TVU-6, 526, TVU-373-1-1, EC-107134, 820 and 920 were found least susceptible. Singh *et al.*, 1985 screened more than 8000 germplasm lines of cowpea for resistance to stored seed damage by *C. maculates* and TVU 2027, TVU 11952 and TVU 11953 were isolated as resistant lines.

Efficacy of oils in the protection of cowpea seeds from *C. maculates*

Application of six different oils viz., Karanz, neem seed, cocunut, groundnut, sawflower, mustard, neem leaves powder and malathion dust were compared in protecting the cow pea seeds from the attack of *C. maculates* Fab. Effectiveness was compared on the basis of the rate of oviposition, population growth and extent of damage. Oils caused high mortality of eggs on the seed surface but had no effect on individuals that successfully entered the seeds. Oil treated seeds were also deterrent to ovipositing females but this activity was limited to a period of 2-4 months. However, malathion was found most effective and could prevent the insect infestation for 8 months.

Caswell (1961, 1970) showed that unthreshed cowpea put into 'rumbus' immediately the crop riped, sustained 24% damage after six months, if left in the field and harvested at a later date. The losses after six months in storage in 'rumbus' was 33%. The damage to threshold cowpea was 87% after nine months of storage. The total loss in storage of cowpea material per annum was calculated to be approximately 7000 tonnes which represents 5% of the 1961/62 production of 8,20,000 tonnes.

All pulses are severely infested by bruchid beetle but cowpea, moong bean and moth bean are comparatively more susceptible to both *Callorobruchus maculetus* and *C. chinesis*. The seeds of cowpea cultivars have been screened for the susceptibility to the attack of pulse beetle but none of them was found to be resistant to pulse beetle attack (Satya Vir, 1981, 1982a, 1982b, 1983d, 1984 and 1985). The seeds were found to be completely damaged by pulse beetle attack in 2-4 month depending on the insect infestation of the seeds. Pulse seeds contain all the four stages viz. eggs, larvae, pupa and adults of *Callorobruchus* species. Thus, a combination of different measures that are together effective against each of these stages will effectively protect grain in storage.

Control

Stored Pulses: The grains should be dried sufficiently to reduce the moisture content as far as possible before storage. The grains may be dried periodically at 2 months interval. Grain should be inspected prior to storing and also at an interval of 2 months after storage. It should be stored preferably in earthen containers with sealed earthen lid or a top cover of sand or ash (7.5 cm thick) or in gunny bags with polythene linning and closed tightly. If there is occurrence of beetle or eggs on seeds are observed in any of the sampling, control measures should be adopted immediately frumigated in an air tight container is recommended with 5 days exposure or heat treatment for 10 min by boiling water or cold treatment at – 12.5°C for 48 hours. Additionally the grain treatment with neem oil or linseed oil 10 ml/kg grain, followed by storage in sealed polythene bags or in jute bags provided the bag surface is sprayed with decis 2.5 ml/m just after storage and again at an interval of 6 months is recommended. None of the treatments affected the germination of cowpea (Rahman, 1991). Pulse grain stored in earthen containers sealed with cowdung, coaltar or gunny bags with polythene lining and mud remained free from external infestation. A cover or a layer of sand over the stored grains in earthen containers also prevented external infestation of pulse beetle (BARI, 1984).

Integrated pest management schedule

Pest management module by making use of agronomic practices, plant extracts and use of resistant varieties to minimize the use of pesticide has been prepared. Early sowing preferably in the last week of June or first or second week of July is most profitable for increasing the crop yield with minimum pest attack in moth bean, moong bean and cow pea crops (Satya Vir and Singh, 1985a). Field trials conducted with various neem extracts revealed that the use of aqueous Neem seed, Neem kernel extracts and Neem seed powder provided complete pest protection umbrella to the kharif crops viz., moth bean, moong bean, cowpea and guar (Satya Vir, 1996 and 2000). Initial application of Neem seed pellets @ 200 kg/ha at the time of sowing was given in furrows for protection of crop from soil pests like termites and white grubs. First spray of Aqueous Extract of Neem seed (NSE) and Neem Seed Kernel (NSKE) extract (2%) were made after 20-30 days of sowing subject to the appearance of insect attack and four such sprays were made at weekly intervals. The results obtained from two years data showed that there was significant reduction in pest attack. The treatment of Neem seed pellets at the time of sowing completely protected the crop from termite attack whereas, plant mortality of 6-12 per cent was recorded in control. Jassid population varied from 12-17 insects per 5 catches in treated plots with Neem extracts compared to 32-44 insects per 5 catches in the control. While fly population ranged 6-9 insects per catch in the treated plots as compared to 22-27 insects per catch in the control. The incidence of yellow mosaic virus (YMV) was also greatly reduced in treated plots with Neem extract. The incidence of Weevils varied from 9-12 insects per 25 sq. meter in treated plots compared to 29-32 insects in the control. Neem extract proved repellent and growth inhibitor against red hairy caterpillar. The larvae stopped feeding and became shrunken in size and died in 3-4 days. The present study, thus, revealed that the use of Aqueous Extract of Neem are antifeedant and repellent to pest attack. These extracts are cost effective as all the raw material is locally available which can be easily exploited by the farmers for its use in the pest management as and when required.

Bhat et al., (1998) also listed various insecticidal formulations of Neem extract for controlling insect pests on cowpea. Pest incidence was lowest in plots treated with monocrotophos at 250 ml/ha and grain yield was highest followed by Neem Seed Extract dose at 25 kg/ha phoselane and quinolphos. Shivan Kar *et al.*, (1998) reported endosulfan at 0.1% as most effective compound and significantly superior to the other treatments in reducing foliage and flower damage and increasing fodder and grain yield in cowpea crop. Ram and Gupta (1989) reported growing the least susceptible variety IGFRI-450, using an optimum fertilizer combination of 80 kg N, 100 kg P_2O_5 and 40 kg K_2O/ha with two weedings at the appropriate stage of the crop growth 15 and 30 days old sowing the crop at the optimum time, the first week of July can give a net profit of Rs. 1477/ha with cost benefit rate of 1: 1: 80.

REFERENCE

Ahmed, R., Lal, S.S., Suresh Chandra, Yadav C.P. and Sachan J.N. 1990. Preliminary field screening for *Aphis Craccivora* resistance in cowpea. Indian Journal of Pulse research 3(1): 77-79.

BARI (Bangladesh Agricultural Research Institute). 1984. Pulse beetle incidence in stored pulses and control measures. Pages 6-13 in Legumes Post harvest technology Project. Annual Report 1983-84. Joydebpur, Gazipur, Bangladesh.

Bhat, N.S., Raju, G.T.T., Manjunatha, M., Nagabhushana, G.G., Deshpande, V.P. 1998. Chemical control of cowpea pod borers. Indian J. of Plant Protection 6(2): 197-200.

Booker, R.H. 1965a. Pests of complex and their control in northern Nigeria. Bull. Ent. Res. 55: 663-672.

Booker, R.H. 1965b. List of insect species found in association with cowpeas at Samaru. Samaru Misc. Paper, No. 9 Institute for Agric. Research Samaru, Ahmadu Bello Univ., Zaria, Nigeria. H pp.

Caswell. G.H. 1961. The infestation of cowpea in the western region of Nigeria. Trop. Sci. 3: 154-158.

Caswell. G.H. 1970. The storage of cowpea in the Northern States of Nigeria. Trop. Sci. 3: 154-158. Samaru Res. Bull. 120. Institute for Agri. Research Samaru, Ahmadu Bello Univ., Zaria, Nigeria.

Gupta, B.N. 1970. studies on chemical control of Galerucid beetle *Madurasia obscurella* Jacoby and jassids, *Empoasca berri* Pruthi on green gram (*Phaseolus aureus* Roxb.). Thesis, P. G. School, Indian Agric. Res. Institute, Delhi.

Jaiswal, P.M. and Petel, S.P. 1993. Comparitive efficacy of various insecticides against *Heliothis armigera* (Hubnu) on cowpea. Indian J. of Entomology 55(3): 339-340.

Jayappa, B.G. and Lingappa, S. 1988. Screening of cowpea germplasm for resistance to *Aphis Craccivora* Koch. in India. Tropical Pest Management. 34(1): 62-64.

Messina, F.J., Renwick, Jaa, Barmore, J.L. 1985. Resistance of *Aphis Craccivora* (Homoptera: Aphidae) in selected varities of cowpea. Journal of Entomological Science (20) 12: 263-269.

Pai, K.M. and Dhuri, A.V. 1991. Effect of insecticides on pest incidence in cowpea. Indian J. of Entomology 53(2): 267-269.

Phelps, R.J. and Oosthuzen, M.J. 1958. Insect injurious to cowpeas in Natal Region. J. Ent. Soc. South Africa. 21: 286-295.

Prevett, P.F. 1961. Field infestation of cowpea (*Vigna Unguiculata)* pods by beetles of the family bruchidae and Curculionidae in Northern Nigeria. Bull. Ent. Res. 52: 635-645.

Rahman, M.M. 1991. Control measures for important insect pests of major pulses. Advances in Pulse Research in Bangladesh. Proc. Of IInd National Workshop on Pulses, 6-8 June, 1989. (Pub. BARI & ICRISAT) Jaydebpur, Bangladesh.

Ram, S., Maurya, R.P. and Gupta, M.P. 1986. Impact of mode of application of insecticides and their dose on the incidence of leaf-hopper (Empoasca berri Pruthu) green fodder and seed yield of summer cowpea. Journal of Ent. Res. 10 (1): 94-98.

Ram, S., and Gupta, M.P. 1989. Integrated Pest Management in fodder cowpea (*Vigna unguiculata* L. Walp.) in India and its economics. Tropical Pest Management. 35(4): 348-357.

Satya Vir. 1980. Oviposition response and development of *Callosobruchus maculates* Fab. On different varieties of cowpea. Bull. Grain Tech. 18 (3): 200-203.

Satya Vir. 1981. Note on the relative susceptibility of different varieties of cowpea to *Callosobruchus maculatus* Fab. Indian J. Agric. Sci. 51 (11): 813-815.

Satya Vir. 1982a Varietal preference in moth (*Vigna aconitifolia* Jacq.) for the pulse beetle *Callosobruchus maculatus*. Fab. Bull. Grain Tech. 20 (1): 3-7.

Satya Vir 1982b Relative susceptibility of cowpea varieties to the pulse beetle *Callosobruchus maculatus*.Pestology, 6 (7): 9-11.

Satya Vir 1983a. Efficacy of some important insecticides for the control of leafhopper *Empoasca berri* Pruthu on moth bean crop. Indian J. of Plant Protection 11: 130-133.

Satya Vir 1983b. Efficacy of some important insecticides against *Amascata moorie* Butler on cowpea crop. Madras Agric. Journal 7: 682-683.

Satya Vir 1983c. Efficacy of some important insecticides for the control of *Bemisia tabaci* Genn., a vector of yellow mosaic disease in mothbean. Indian J. of Plant Protection 1: 31-33.

Satya Vir. 1983d. Varietal resistance and susceptibility of cowpea to *Callosobruchus maculatus* Fab. Indian J. Ent. 45 (3): 213-217.

Satya Vir. 1983e. Assessment of yield loss in moth bean and cowpea crops due to insect pests and their control. Proc. 10th International Congress Plant Prot., England 6F. Vol. 3:3.

Satya Vir. 1984. Screening of some cowpea varieties against the pulse beetle *Callosobruchus chinensiuss* Linn. Seeds and Farms 9 (6): 29-31.

Satya Vir. 1985. Studies on the bionomics of *Bemisia tabaci*, a vector of yellow mosaic virus, extent of damage and the control in mothbean. Proc. Adv. In Develop. Bio. Of Animals. Eds: S.C. Goyal and C.B.L. Srivastáva. ISDB, Poona pp: 145-148.

Satya Vir. 1989. Multiplication of *Callosobruchus maculatus* Fab. Infestation and loss in weight of mothbean and cowpea seeds at different level of infestation. Bull. Grain Tech. 27(1): 21-25.

Satya Vir. 1996. Neem-Prtential Pesticide. Pest Management 20 (5): 23-28.

Satya Vir. 1998. Studies on the bionomics of *Bemisia tabac* Gen. (Aleyrodidae)- Avector of yellow mosaic virus, extent of damage, pest resistance and their management in mothbean. Proc. 2nd

Inter. Workshop on *Bemisia* and *Germini Virus* and their control. June 7-11, 1998. Puerto Rica, U.S.A.

Satya Vir 2000. Insect Pest Management in leguminious crops of Thar desert with Neem products. Proceeding International Conference of Neem. May 19-21, 1999. University of British Colombia, Vancouver, Canada.

Satya Vir and Jindal, S.K. 1981. The oviposition and development of *Callosobruchus maculatus* Fabricius (Coleoptera: Bruchidae) on different host species. Bull. Grain Tech. 19 (3): 180-189.

Satya Vir and Jindal, S.K. 1982. Assessment of yield loss due to insect pest of pulses in cowpea, *Vigna unguiculata* (L) Walp. Seeds and Farms. 8 (8 & 7): 59-60.

Satya Vir and Singh, M. 1985a. Influence of date of sowing on incidence of and extent of damage of insect pests in moth bean and cowpea. Ann. Arid Zone. 24 (4): 329-333.

Satya Vir and Singh M.P. 1985b. Effect of different host plants on growth and development of *Amscata moorie* Rec. Adv. Devel. Biology of Animals. Eds: S.C. Goyal and C.B.L. Shrivastava ISCB, Poona. pp: 149-152.

Saxena, H.P. 1978. Pest of grain legumes and their control in India. In pest of grain Legumes: Ecology and Control, Eds: Singh, Emden and Taylor. Academic Press pp: 15-24.

Saxena, H.P., Phokela, A. and Singh Y. 1975. Effectiveness of Carbofuran as seed treatment for controlling pest complex of 'moong' (*Phaseolus aureous)* and 'Urd' (*Phaseolus mungo)*. Ent. News. 5: 27-28.

Saxena, H.P., Sircar, P. and Phokela, A. 1970. Predation of *Coccinella septenpunctata* Linn. *Ischiodon scutellaris* Fab. On *Aphis craccivora* Koch. Indian J. Ent. 32: 105-106.

Shivan Kar, V.J., Ram, S. and Gupta, M.P. 1998. Extent of damage by insect pests pand crop yield as influenced by insecticidal treatments in fodder cowpea (*Vigna unguiculate* L. Walp.). Indian J. of Plant Protection 16 (2): 221-224.

Singh, B.B., Singh S.R. and Adjadi, O. 1985. Bruchid resistance in cowpea. Crop Science 25 (5): 736-739.

Taylor, T.A. 1964. The field pest problems on cowpea, *Vigna sinesis* L. in Southern Nigegia. Nig. Grower Producer 3: 1-4.

CHAPTER - 11

COWPEA NEMATOLOGY

R.K. Kaul

ABSTRACT

Surveys of cowpea crop in different states of India have revealed existence of large number of plant nematode species on this crop. Amongst them, *Heterodera cajani, Meloidogyne incognita* and *Rotylenchulus reniformis* are considered to be economically important species. Effects of these nematode species either alone or in combination with other nematodes or pathogenic fungi or bacteria on growth of plants and their multiplication have been studied by many researchers. Large number of cowpea varieties have also been screened for resistance against different nematodes. Genotype Co-1, Co-3, V-16 and K 4967 have been recorded resistant to *R. reniformis* and *M. incognita.* Phenamiphos seed treatments of cowpea or seed soaking in 0.1% technical a.i. of carbofuran, dimethioate, methyl demetone, monocrotophos, phosphamidon or triazophos for 12 hours have been observed to reduce *M. incognita* gall index of roots. Similarly, *Pasteuria penetrans* has been observed to reduce number of *H. cajani* cysts on cowpea.

INTRODUCTION

Cowpea is an important crop grown in tropic and sub tropic zone of India. The crop is known to support large number of plant parasitic nematodes. Surveys of cowpea crop in different states of India including Utter Pradesh, Haryana,

Punjab, Rajasthan, Maharashtra, Kerala, Himachal Pradesh, Orissa, Tamil Nadu have shown existence of one species or the other on this crop. Various plant nematode species reported on this crop are *Meloidogyne incognita, M. javanica, M. arenaria, M. hapla, Pratylenchus zeae, Heterodera cajani, Psilenchus hilarulus, Rotylenchulus reniformis, Paratrichodurus mirzae, Aphelenchus radicicola, Tylenchorhynchus goffarti, T. mashoodi* and *Helicotylenchus abbunnamai* (Dass *et. al.*1979, Kaushik and Bajaj, 1980, Hassan *et. al.* 1998). Amongst various nematode species reported on cowpea, *H. cajani, M. incognita* and *R. reniformis* are considered to be economically important.

Nematode life cycle and multiplication studies on cowpea

Life cycle and multiplication of nematode on cowpea have been studied by many researchers. Edward and Mishra (1968) described life cycle of *H. cajani* (then *H. vigni*) on cowpea. Later Gupta and Edwards (1973) described various developmental stages in the life cycle of this nematode. They observed 3rd stage after 5 days and 4th stage after 11 days. Males were observed to complete life cycle in 13 days while females in 17 days. Biology of *H. cajani* on cowpea along with other plants was studied by Yadav and Walia (1988). They recorded complete life cycle of the nematode in 29 days on cowpea at temperatures ranging between 24.3 to 38.8°C. Sharma and Sethi in 1975 studied the effect of initial inoculum levels of *H. cajani* on the growth of cowpea and multiplication of the nematode. They observed maximum population and rate of multiplication of nematode at initial inoculum level of 100 larvae and 10 larvae, respectively. Similarly, multiplication of *Rotylenchulus reniformis* on susceptible and resistant varieties of cowpea was studied by Haque and Padmavathy (1985). They observed final population of the nematode to be maximum at an initial inoculum level of 10,000 larvae while highest rate of multiplication at 100 larvae. Panda and Sheshadri (1979) studied the effect of initial inoculum level of *R. reniformis* on the growth of cowpea. They observed cowpea to be susceptible to the nematode and some of the growth attributes of the plant were reduced significantly at an initial inoculum level of 100 larvae/kg soil. Khan and Saxena (1992) studied the effect of different inoculum levels of *H. cajani* on the growth of cowpea.

They observed reduction in plant growth at 100 larvae or above level of *H. cajani*.

Interaction of nematode with other biotic agents on cowpea

Interactions between nematode-nematode and nematode-fungus and nematode-bacteria have been studied by many researchers. Sharma and Sethi (1976a) studied the interaction between *M. incognita* and *H. cajani* singly and in combination on the growth of cowpea. They observed significant reduction in the growth of cowpea when both the nematodes were inoculated either singly or in combination. They also observed that *M. incognita* infection reduced nitrogen content of shoot. Similarly, Sharma and Sethi (1976b) analysed root nodules for leghaemoglobin content and observed *M. incognita* infection to cause more reduction in leghaemoglobin content than *H. cajani* infection. Walia and Gupta (1986a,b) studied the interaction between *H. cajani* and *Rhizoctonia solani* on cowpea. They observed *R. solani* to adversely affect *H. cajani* population when inoculated one week prior to nematode. Inoculation of the fungus two weeks after the nematode was observed to reduce top growth of the plant significantly than compared to check and other treatments. Varaprasad *et. al.* (1987) studied the effect of single and concomitant inoculation of *H. cajani* and *Fusarium solani* on cowpea growth. *H. cajani* when inoculated alone was observed to decrease shoot weight and leaf area than compared to their concomitant inoculation. Khan and Hussan (1988) studied the effect of individual, concomitant and sequential inoculation of *Rhizobium*, *R. reniformis*, *M. incognita* and *Rhizoctonia solani* on cowpea disease development and nematode multiplication. Amongst different pathogens tested, they observed *Rhizoctonia* to be the most aggressive pathogen. This was followed by *M. incognita* and *R. reniformis*. Concomitant inoculation of nematode and fungus was observed to be more damaging then the association of both the nematode species but *M. incognita* in association with *R. solani* caused greater plant growth reduction then its association with *R. reniformis*. They also observed more reduction in the growth of cowpea when inoculated alone than in combination with *Rhizobium*. Prior inoculation of the *Rhizobium* by fifteen days to *M. incognita* or *R. reniformis* was observed to reduce nematode multiplication, root-knot index and improved nodulation of the roots. In sequential inoculations, interaction was found to be more dependent. Prior establishment of either nematode species was invariably observed to be antagonistic to the multiplication of the other subsequently inoculated species. *R. solani* whether inoculated simultaneously or sequentially reduced the rate of multiplication of both the nematode species than compared to alone. Varaprasad and Kumar (1991) studied the effects of single and combined inoculation of *H. cajani* and *F. solani* on activity of nitrate reductase and glutamine synthetase in cowpea. They recorded reduction in both the enzymes when *H. cajani* and *F. solani* were inoculated alone than their combined inoculation.

Influence of mycorrhizal fungus on host parasite relationship on cowpea has also been studied by many researchers. Jain and Sethi (1988a) studied the influence of G. *fasciculatum* and *G. epigeous* on the penetration and development of *H. cajani* on cowpea. Establishment of mycorrhizal fungus by 15 days prior to the nematode was observed to adversely affect root-knot nematode penetration then their simultaneous inoculation. More than 60 per cent colonization of root system by VAM was observed to adversely affect penetration of the nematode in the roots. However, once nematode invaded the roots, their development was not found to be hampered by the fungus. Jain and Sethi (1988b) also studied the effect of the time of application of *G. fasciculatum* on *M. incognita* and *H. cajani*. Prior establishment of VAM fungus by one or two weeks was found to mitigate the adverse effect of *M. incognita* and *H. cajani* on cowpea growth. Presence of either nematode was observed to affect mycorrhizal root infection and spore production. Lingaraju and Gowsami (1993) studied the interactive relationship between *G. fasciculatum* and *R. reniformis* on cowpea under green house conditions. Mycorrhizal fungus was observed' to enhance total biomass and fresh weight of root while nematode was observed to reduce the same. The final population of nematode was recorded to be more in the presence of the fungus. The nematode was not observed to influence the spore numbers though it interfered with the mychorrhizal colonization. The

VAM was observed to increase tolerance in cowpea to the nematode even in the presence of damaging levels of nematode. Santhi *et. al.* (1993) also studied the effect of *G. fasciculatum* and *M. incognita* on cowpea under different moisture levels. They recorded maximum growth of plants at 70 % moisture level which decreased with further increase in the moisture level. In case of root height, however, reverse trend was noticed. They also observed a positive relationship between nematode population and root colonization by the fungus with moisture. Similarly, Sivaprasad *et. al.* (1990) studied the effect of combined inoculation of *Rhizobium*, VAM fungus and *M. incognita* on growth of cowpea. They recorded significant increase in the growth of cowpea and nodulation when inoculated with *Rhizobium* and VAM fungi. Inoculation of VAM was observed to reduce *M. incognita* multiplication.

Biochemical changes induced as a result of nematode infection have also been studied by many workers. Sharma and Sethi (1976a) reported increase in non-reducing sugars in the shoot as a result of *M. incognita* infection in tested varieties of cowpea. Infection of *H. cajani* on the other hand led to the decrease in reducing sugars in the shoots of both the varieties (i.e. Pusa Barsati and Barsati ,mutant) while non-reducing sugars increased in shoots and decreased in the roots of plants infected with *H. cajani*. The non-reducing sugars accumulated more in *H. cajani* affected shoots then in *M. incognita* infected plants suggesting there by a relatively more interference in translation by *H. cajani* then by *M. incognita*. In general, total carbohyderate content was also observed to increase as a result of nematode infection which was more in case of *H. cajani*. Sebastian *et. al.* (1992) studied variation in tyrosine activity as a result of *M. incognita* infection in cowpea. The enzyme activity was observed to be enhanced at all stages of infection. Mohanty *et. al.* (1988) observed that isonezyme appearing after infection of the nematode differed considerably with regards to different varieties. Similarly, Ganguly and Dasgupta (1988) also observed increased superoxidase dismutase activity in susceptible Pusa Barsati variety as compared to C152 82-1-B (resistant variety). Sirohi and Dasgupta (1993 a,b) investigated the mechanism of resistance in cowpea against *M. incognita* infection. They observed higher activity in phenylalanine ammonia lyase in inoculated resistant cowpea cultivar C-152 after 24-72 hrs. of nematode inoculation. They demonstrated that cholorogenic acid (CGA) turn over values were higher in inoculated resistant varieties in the period ranging between 24-72 hrs. They further reported that root-knot nematode infection in the host tissue resulted in the depression of noval genes which subsequently lead to the de novo synthesis of RNA and protein molecules under the influence of illicitors released by the parasite.

Screening of cowpea germplasm for nematode resistance

Screening of cowpea genotypes, against nematode pathogen has also been done by many researchers. Darekar and Patel (1981) screened 13 varieties of cowpea against *M. incognita.* Though all varieties exhibited galling but two varieties C-152 and 82-1-B expressed some degree of resistance. Thakur and Patel (1984) screened six varieties of cowpea against *R. reniformis* and found V-16 variety to be resistant while Pusa Phalguni as moderately resistant. Haque and Padmavathy (1985) reported significantly reduced multiplication of *R. reniformis* on C-152 cultivar. Jain and Bhati (1986) reported results of screening of sixty two varieties/lines of cowpea against *M. javanica.* None of the tested varieties showed resistance. However, C-21, GC-11, G-17m and GC-3 showed average gall index of 2-3 and were categorized as moderately resistant. Patel *et al.* (1988) screened sixty cowpea lines in microplots having heavy infestation of *M. incognita* and *M. javanica.* None of the tested lines were observed to be resistant to the nematode. Only some lines exhibited moderate resistance. Devi and Gupta (1991) screened fifteen cowpea strains for resistance against *H. cajani* and observed only two Orissa red and IFG 8505 to be resistant. Balasubramanian *et al.* (1996) screened 45 cowpea lines for resistance against *H. cajani* and reported only two cultivars, Medium 11912 and RT 11936 to be moderately resistant. Subramaniyan *et al.* (1997) also screened 37 genotypes of cowpea against root-knot and reniform nematode. Genotype Co-1, Co-3 and K4967 were observed resistant to the nematode.

Nematode management on cowpea

Many studies have also been conducted on the management of the pathogenic nematode species on cowpea. Reddy (1984) studied the efficacy of seed treatment with the nematicide for control of *M. incognita* on cowpea. He observed lowest root-knot index in cowpea in phenamiphos seed treatment. Verma and Gupta (1986) studied the effect of NPK fertilizer singly and in combination with the pesticides, aldicarb and carbofuran at different dosages on *M. incognita* infection. They recorded significant increase in plant growth and minimum number of egg masses in treatment having 1000 larvae along with 1 kg a.i./ha aldicarb. Walia and Gupta (1986a) studied the effect of *Rhizoctonia bataticola* on population of *H. cajani* on cowpea. They observed adverse effect on nematode population by the fungus with maximum reduction being observed when fungus established seven days prior to nematode inoculation. Mehr and Sethi (1988) reported suppressing effect of phenamiphos on egg hatching of *M. incognita*. They also reported exposure of larvae to sublethal concentration of phenamiphos for 24 and 48 hours which caused significant reduction in the penetration of the larvae. Verma and Gupta (1994) observed lowest root-knot index up to 30 days when the cowpea seeds were soaked in 0.1% technical a.i. of carbofuran, dimethioate, methyl demetone, monocrotophos, phosphamidon and triazophos for 12 hours. Mehr *et. al.* (1995) reported marked reduction in the population of ecto and migratory endoparasitic nematodes infecting cowpea with the furrow application of sebuphos @10kg/ha in three cropping seasons. Singh and Dhawan (1996) investigated extent of suppression of *H. cajani* by *Pasteuria penetrans* during three successive plantings of cowpea. They recorded significant reduction in the number of cysts and juveniles in the treatment receiving combined inoculation of *H. cajani* and *P. penetrans* compared to treatment receiving *H. cajani* alone. Singh and Dhawan (1998) tested the compatibility of a new strain of *Pasteuria penetrans* with a nematicide on the management of *H. cajani* on cowpea. They observed a significant increase in plant growth when carbofuran was combined with *P. penetrans* than carbofuran alone. Similar trends were observed in the reduction of juvenile population in the soil. The study also indicated that the nematode had no adverse effect on the growth of the bacteria. Mojumdar and Mittal (1999) studied the effect of storage after coating the seeds of cowpea with neem products against *H. cajani.* They observed neem treatment to seeds had no adverse effect on the germination of seeds even after three months of storage and plants from such seeds showed reduced infection to *H. cajani.* They observed powdered neem seed and neem cake to be the most effective product. Mohan and Dhawan (2000) compared penetration of *H. cajani* larvae parasitized by 20, 40 and 60 spores of *P. penetrans* in cowpea cv. Pusa Komal. They observed reduced percent penetration of larvae with the increase in the number of spores encumbering them. Devi (2000) studied the efficacy of neem bark, cake, leaf and flower ash (each at 500mg/kg soil) and commercial product of neem at 2 and 4% as seed treatment against *H. cajani* on cowpea. They recorded lowest number of cysts from neem bark followed by seed and leaf powder.

REFERENCES

Balasubramanian, P., Vadivelu, S. and Vijaykumar, J. (1996). Reaction of cowpea varieties against *Heterodera cajani* Koshy. *Indian J. Nematol.* 9: 99-100.

Darekar, K.S. and Patel, B.D. (1981). Reaction of some cowpea varieties to root-knot nematode. *Indian J. Nematol.* 11: 83-84.

Das, S.N., Mishra, C.D. and Mohanty, K.C. (1979). Occurrence and host preference of some plant parasitic nematodes on pulse crops. *Indian J. Nematol.* 9:74.

Devi, L.S. and Gupta, P. (1991). Evaluation of cowpea varieties against cyst nematode. *Current Nematology. 2: 59-60.*

Devi, L.S. (2000). Effect of neem products against *Heterodera cajani* on pigeon pea and cowpea. *Indian J. Nematol.* 30: 250-252.

Ganguly, A.K. and Dasgupta, D.R. (1988). Superoxide dismutase (E.C. 1.15.1.1) activity in resistant and susceptible cowpea cultivars inoculated with root-knot nematode, *Meloidogyne incognita. Indian J. Nematol.* 18: 322-325.

Gupta, Pramila and Edward, J.C. (1973). Studies on the biology of *Heterodera vigni* (Heteroderidae: Nematoda) I – life cycle. *Indian J. Nematol.* 3: 99-108.

Haque, M.M. and Padmavathi, P.V. (1985). On multiplication of *Rotylenchulus reniformis* in susceptible and resistant cultivars of cowpea. *Indian J. Nematol.* 15: 223-225.

Hassan, N., Pandey, K.C. and Hazra, C.R. (1998). Plant parasitic nematodes associated with major fodder crops in India. *Indian J. Nematol.* 28: 140-144.

Jain, R.K. and Bhatti, D.S. (1986). Screening of cowpea germplasm againsty root-knot nematode, *Meloidogyne javanica. Indian J. Nematol.* 16: 113-114.

Jain, R.K. and Hasan, N. (1986). Efficacy of neem cake on fodder production, photosynthetic pigments and nematodes associated oats and its residual effect on cowpea. *Indian J. Nematol.* 16: 98-100.

Jain, R.K. and Sethi, C.L. (1988a). Interaction between vesicular arbuscular mycorrhiza and *Meloidogyne incognita, Heterodera cajani* on cowpea as influenced by time of inoculation. *Indian J. Nematol.* 18: 263-268.

Jain, R.K. and Sethi, C.L. (1988b). Influence of endomycorrhizal fungi *Glomus fasciculatum* and *G.epigaecus* on penetration and development of *Heterodera cajani* on cowpea. *Indian J. Nematol.* 18: 89-93.

Kaushik, H.D. and Bajaj, H.K. (1980). Nematodes associated with pulse crops in Haryana. *Indian J. Nematol.* 10: 92-96.

Khan, T.A. and Saxena, S.K. (1992). Effect of inoculating cowpea with different inoculum levels of *Heterodera cajani. Afro-Asian Journal of Nematology Network. 1: 18-20.*

Khan, Tarbeiz Ahmad and Hussain, Syed Israr (1988). Effect of individual, concomitant and sequential inoculation of *Rhizobium, Rotylenchulus reniformis, Meloidogyne incognita* and *Rhizoctonia solani* on cowpea plant growth, disease development and nematode multiplication. *Indian J. Nematol.* 18: 232-238.

Lingaraju, S. and Goswami, B.K. (1993). Influence of a mycorrhizal fungus *Glomus fasciculatum* on the host-parasite relationship of *Rotylenchulus reniformis* in cowpea. *Indian J. Nematol.* 23; 137-141.

Meher, H.C. and Sethi, C.L. (1988). Effect of phenemiphos on hatching and invasion of *Meloidogyne incognita* on cowpea. *Indian J. Nematol.* 18: 186-191.

Meher, H.C., Nand, S., Kaushal, K.K. and Singh, G. (1995). Effect of sebuphos on soil nematodes following former application to cowpea, okra, pea, soybean and tomato. *Indian J. Nematol.* 25: 88-93.

Mohan, S. and Dhawan, S.C. (2000). Penetration and biochemical changes induced in *Pasteuria* infected *Heterodera cajani* in cowpea roots. *Indian J. Nematol.* 30: 141-146.

Mohanty, K.C., Ganguly, A.K. and Dasgupta, D.R. (1988). Partial purification and characterization of peroxidase (E.C.1.11.1.7) from two varieties of cowpea, inoculated with root-knot nematode *Meloidogyne incognita. Indian J. Nematol.* 18: 252-262.

Mojumdar, V. and Mittal, A. (1999). Effect of storage after seed coating of cowpea with neem products against *Heterodera cajani. Pesticide Research Journal. 11: 188-189.*

Panda, M and Seshadri, A.R. (1979). Effect of initial inoculum level of *Rotylenchulus reniformis* on different growth attributes of cowpea, mungbean and urd. *Indian J. Nematol.* 9: 59-60.

Patel, B.A., Chavda, J.C., Patel, H.V. and Patel, D.J. (1988). Screening of certain cowpea lines for resistance to root-knot nematodes (*Meloidogyne incognita* and *M. javanica*). *Indian J. Nematol.* 18:128.

Patel, D.J., Patel, H.V. and Patel, S.K. (1993). Estimation of avoidable yield losses in green gram and cowpea due to root-knot nematodes. *Indian J. Nematol.* 23: 213-224.

Patel, G.H., Shah, H.M. and Patel, D.J. (1977). Screening of cowpea cultivars against root-knot nematodes. *Indian J. Nematol.* 7: 169-170.

Reddy, P. Parvatha (1984). Efficacy of seed treatment with three nematicides for the control of *Meloidogyne incognita* infecting cowpea., frenchbean and peas. *Indian J. Nematol.* 14: 39-40.

Santhi, A. and Sundarababu, Rajeshwari. (1993). effect of different levels of moisture on the interaction of *Glomus fasciculatum* with *Meloidogyne incognita* on cowpea. *Indian J. Nematol.* 23: 199-201.

Sebastian, E., Nidiry, J., Chandravandana, M.V. Khan, R.M. and Reddy, P.P. (1992). Variations induced by *Meloidogyne incognita* in tyrosinase activity in the roots of resistant and susceptible varieties of cowpea and tomato. *Indian J. Nematol.* 22: 11-13.

Sharma, N.K. and Sethi, C.L. (1975). Leghaemoglobin content of cowpea nodules as influenced by *Meloidogyne incognita* and *Heterodera cajani. Indian J.Nematol.* 5: 113-114.

Sharma, N.K. and Sethi, C.L. (1976a). Influence of *Meloidogyne incognita* and *Heterodera cajani* on carbohydrate content of cowpea. *Indian J. Nematol.* 6: 171-173.

Sharma, N.K. and Sethi, C.L. (1976b). Interrelationship between *Meloidogyne incognita, Heterodera cajani* and *Rhizobium* sp. on cowpea (*Vigna sinensis* (L)San's*). Indian J. Nematol.* 6: 117-123.

Singh, Bansa and Dhawan, S.C. (1996). Suppression of *Heterodera cajani* by *Pasteuria penetrans* during three successive plantings of cowpea. *Indian J. Nematol.* 26: 216-221.

Singh, Bansa and Dhawan, S.C. (1998). Compatibility of *Pasteuria penetrans* with carbofuran in controlling *Heterodera cajani* on cowpea. *Indian J. Nematol.* 28: 110-113.

Singh, D.B. and Reddy, P. Parvatha (1986). Inheritence of resistance to root-knot nematode in cowpea. *Indian J. Nematol.* 16: 284-285.

Sirohi, Anil and Dasgupta, D.R. (1993a). Mechanism of resistance in cowpea to the root-knot nematode, *Meloidogyne incognita* race 1:1 Early induction of Phenylalanine ammonia Lyase (E.C. 4.3-1.5) and chlorogenic acid. *Indian J. Nematol.* 23: 31-41.

Sirohi, Anil and Dasgupta, D.R. (1993b). Mechanism of resistance in cowpea to the root-knot nematode, *Meloidogyne incognita* race I: II : De novo synthesis. *Indian J. Nematol.* 23: 45-52.

Sivaprasad, P., Jacob, Arthur and George, Babu (1990). Root-knot nematode infestation and nodulation as influenced by VA mycorrhizal association in cowpea. *Indian J. Nematol.* 20: 49-52.

Subryamaniyan, S., Faziullah, A.K., Rajendran, G. and Vadivelu, Sivagam. (1997). Reaction of some mung, cowpea, lablab and tomato genotypes to the reniform and root-knot nematode. *Indian J. Nematol.* 27: 130-131.

Thakar, N.A. and Patel, C.C. (1984). Screening of few cowpea varieties against reniform nematode, *Rotylenchulus reniformis* Linford and Oliveira, 1940. *Indian J. Nematol.* 14: 204.

Varaprasad, K.S., Kumar, P.A. (1991). Effect of single and concomitant inoculations of *Heterodera cajani* and *Fusarium solani* on activities of nitrogen reduction and glutamine synthetase in cowpea. *Nematologia Mediterranea. 19: 55-56.*

Varaprasad, K.S., Polisetty, R. and Sethi, C.L. (1987). Effect of separate and concomitant inoculation of *Heterodera cajani* Koshy, 1967 and *Fusarium solani* Mart (Sacc.) on cowpea growth and photosynthetic rate. *Tropical Grain Legume Bulletin. 34: 29-31.*

Verma, K.K. and Gupta, D.C. (1994). Management of *Meloidogyne javanica* on cowpea (*Vigna unguiculata*) with seed treatment and evaluation of its genotype. *Indian J. Nematol.* 24: 234-235.

Verma, K.K. and Gupta, K.K. (1986). Studies on effect of NPK fertilizers singly and in combination with pesticides on cowpea (*Vigna unguiculata* (L) Walp.) infected with *Meloidogyne javanica. Indian J. Nematol.* 16: 51-55.

Walia, Kum Kum and Gupta, D.C. (1986a). Effect of the fungus, *Rhizoctonia bataticola* on the population development of pigeon pea cyst nematode, *Heterodera cajani* on cowpea (*Vigna unguiculata*). *Indian J. Nematol.* 16:131-132.

Walia, Kum Kum and Gupta, D.C. (1986b). Antagonism between *Heterodera cajani* and *Rhizoctonia solani* on cowpea (*Vigna unguiculata* (L) Walp.). *Indian J. Nematol.* 16: 41-43.

Yadav, U.S. and Walia, R.K. (1988). On biology of pigeon pea cyst nematode, *Heterodera cajani* Koshy 1967. *Indian J. Nematol.* 18: 35-39.

CHAPTER - 12

COWPEA FORAGE ASPECTS

C.R. Ramesh, P.S. Pathak and B. Gangaiah

ABSTRACT

Considerable variation exists among cowpea accessions, indicating the potential for selection of superior and better adapted genotypes either for forage or dual purpose (grain – forage) use. Many cultivars have been recommended to different agroclimatic zones, for intensive, relay and overlapping cropping systems in the country. Cultivars which are dual purpose can be recommended for crop livestock farmers because they contribute both to human food and have substantial forage yields. Cultivation of dualpurpose accessions benefits poor farmers through improved forage yield without requiring additional land or inputs (seed, fertilizer, labour). Adoption of dual purpose types with high grain production, increased the quantity and quality of crop residue and high leaf to stem ratio, would provide the most benefit for poor small-holder farmers in India. Cowpea can be used at all stages of growth as a vegetable crop. The tender green leaves are an important food source in some parts of India and are prepared as a pot herb, like spinach. Immature snapped pods are used in the same way as snap beans, often being mixed with other foods. Green cowpea seeds are boiled as a fresh vegetable, or may be canned or frozen. Dry mature seeds are also suitable for boiling and canning.

Increased fodder from cowpea crop residues could enhance livestock production substantially without requiring additional inputs from resource-poor livestock farmers. India has been facing frequent drought situations in the recent past. Adoption of dual purpose cowpea would go a long way in serving as protein source to the resource poor and their livestock

CROPS

The demand for livestock products like meat and milk in developing countries is expected to double by the year 2020 as a result of improvement in life styles due to rising incomes, and increased urbanization (Delgado *et al.*, 1999). The majority of livestock production in India is by small-holder systems, characterized by low milk production growth rates, extended calving and lambing intervals. They depend mainly on local forages mostly comprised of post-grain harvesting crop residues and pastures that are high in fiber and deficient in many important nutrients. Leguminous forages or food-feed crops, which are rich in protein and minerals have the ability to enhance fiber degradation in the rumen hence, can serve as strategic supplements to other forage resources. Cowpea grain contains about 25% protein and 64% carbohydrate (Bressani, 1985) and therefore, has a tremendous potential to contribute to alleviation of malnutrition among resource-poor farmers. In addition, cowpea contributes to the sustainability of cropping systems and soil fertility improvements in marginal lands by providing ground cover and plant residues, fixing nitrogen and suppressing weeds. Some cowpea varieties also cause suicidal germination of *Striga hermonthica*, a devastating parasitic weed of cereals. Hence, cowpea is an ideal crop for the semi-arid regions of the tropics, where other food legumes may not perform well. A number of varieties have been developed in IITA, Ibadan combining diverse plant types, different maturity periods, and resistance to several diseases, insect pests and parasitic weeds and possessing other good agronomic traits (Singh *et al.*, 1997)

Vigna unguiculata (L.) Walp is grown in over two-thirds of the developing world, usually as a companion or relay crop with small grains or corn (*Zea mays* L.). Development of new varieties resistant to insects and pests or have shorter life cycles has contributed to increased cultivation of this crop. Cowpea is adapted to warm weather and requires less rainfall than most crops, therefore it is primarily cultivated in the semi-arid regions of the lowland tropics and subtropics, where soils are poor and rainfall is limited (Mortimore *et al.*, 1997). The use of cowpea as fodder is attractive in mixed crop/ livestock systems where both grain and fodder can be obtained from the same crop (Tarawali *et al.*, 1997). Tarawali *et al.* (1997) has reported that cowpea grain typically contains 230-250g/kg crude protein (CP) and 500-670 g/kg starch on a dry matter (DM) basis; and cowpea forage, i.e. the crop residue contains almost, 210g CP and 600g digestible dry matter per kg DM. There is increasing emphasis on integrating crop livestock production to promote more sustainable agricultural systems. Cowpea can make a very important contribution towards livestock fodder and supply N to the soil (Lat *et al.*, 1978). The forage is used as a ruminant feed by smallholder farmers in West Africa, Asia and South America therefore, offers potential for use in the drier regions of India. Under such conditions cowpea forage is usually superior to other forage legumes in terms of both quantity and quality. The use of cowpea as fodder is most advanced in India, where green crop is used for grazing or cut and mixed in combination with cereals and other crops for lactating cows. Cowpea in India can maintain milk yields of >1.5 gallon cow^{-1} day^{-1}. Inclusion of green cowpea pods in the fodder maintains nutritive value. Trials on fodder varieties of cowpea in India gave dry-matter yields of >1.8 ton $acre^{-1}$ and protein contents up to 26% (Relwani *et al.*, 1970). Cutting trials have indicated that harvesting of cowpea 60 days after planting, provided the best dry matter yields of highest quality (Kandaswamy *et al.*, 1976). Grain cowpea in semi arid and arid zones of West Africa is planted between pearl millet rows 2 to 3 weeks after planting of the latter, followed by fodder cowpea 3 to 4 weeks later. Following harvest of pearl millet, the grain cowpea is harvested and the fodder cowpea is left to grow. Typical yields obtained are 350 to 400 pounds dry cowpea fodder $acre^{-1}$ (Singh, 1993). Under appropriate management, cowpea can provide good quality fodder for *in situ* grazing, silage (in combination with cereals) or hay. The management and selection of cultivars will depend on the farming system requirements and the mode of use (Tarawali *et al.*, 1997). Cowpea will provide a crucial role as it facilitates crop-livestock integration, which is associated with intensif-ication and land conserving investments. In fixing N, cowpea also brings this plant nutrient into the cycle. Its economic function in the system is complementary to that of cereals (Mortimore *et al.*, 1997). A best management practice is, it reduces soil erosion and conserves water, while at the same time it must increase land productivity and conserving fuel is, therefore, a conservation tillage. Any tillage and planting system that covers 30% or more of the soil surface with crop residue after planting is considered conservation tillage (Gallaher and Hawf, 1997)

Cowpea is an annual erect or semi-erect or spreading C3 legume with quick and high biomass production capacity. It is cultivated for dry grain, fresh beans, edible pods, edible leaves, hay /forage and green manuring purposes. In North India it is grown during the monsoon and summer seasons while in South it is grown year round. The area under fodder cowpea in the country is around 0.3 m ha (Hazra and Sinha, 1996).

Climate and soils

Cowpea is a crop of tropics and subtropics with wide distribution extending from peninsular region in south to Lower Himalayas in the North up to 1500 m AMSL, however, it is also grown in moderately humid regions of the country. The optimum temperature for growth required is 23-28°C, the crop can also tolerate high temperatures up to 45°C. High night temperatures may however, reduce vegetative growth and bring early flowering (Grantz and Hall, 1982). Further high soil temperatures (35°C) reduces seedling emergence that can be overcome by shallow (2.5 cm) sowing (Warang and Hall, 1984a). Chilling also reduces the crop establishment as crop requires average soil temperatures of >19°C for 3 days from sowing to emergence (Hall, 1999) that limits its cultivation during *rabi* season in North. Thus, too high and too cold temperatures are not suitable for crop establishment.

Cowpea is adapted in wider soils, ranging from sandy to clays. Good drainage is primary soil requirement and thus, waterlogged soils are unfit for cowpea culture. The crop performs better in deep, well drained, fertile and neutral soils. Cowpea is moderately susceptible to soil salinity with an initial yield declines at EC of 1.3 dS/m and beyond this, yield declines considerably with increasing salinity (Moss and Hoffman, 1977). The crop is resistant to drought, low soil fertility and acidity (Donald *et al.*, 1991).

Growth Habits

Plant types are often categorized as erect, semi-erect, prostrate (trailing) or climbing. There is much variability within the species. Growth habit ranges from indeterminate to fairly determinate with the non-viny types tending to be more determinate. Cowpea generally is strongly tap rooted. Root depth has been measured to 95 inches in 8 weeks after seeding.

The seed coat can be either smooth or wrinkled and of various colors including white, cream, green, buff, red, brown and black. Seed may also be speckled, mottled, or blotchy. Many are also referred to as "eyed" (black-eyed, pinkeye purple hull, etc.) where white colored hilum is surrounded by another color.

Emergence is epigeal (similar to common bean and lupin) where the cotyledons emerge from the ground during germination. This type of emergence makes cowpea more susceptible to seedling injury, since the plant does not regenerate buds below the cotyledonary node.

The cowpea seed protein is rich in the amino acid contents (lysine and tryptophan), compared to cereal grains; however, it is deficient in methionine and cystine when compared to animal proteins. Therefore, cowpea seed is valued as a nutritional supplement to cereals and an extender of animal proteins.

Development of dry-season dual-purpose cowpea varieties

In the semi-arid regions, cowpea fodder is an important resource for livestock. The above ground parts of cowpea, except pods, are harvested for fodder. The take-off of cowpea fodder contributes to feed supplies for large and small ruminants. Traditionally, farmers cultivate two main types of cowpea: early maturing varieties grown for grain and late maturing varieties grown for fodder production. However, during the dry season, good quality fodder is scarce.

Development of dual-purpose varieties that would give reasonable grain and fodder yields would go a long way in maximizing the output from land and labor. The use of cowpea as a dual-purpose crop, providing both grain and fodder, is attractive in mixed crop/livestock systems where land and feed are becoming increasingly scarce (Tarawali *et al.*, 1997) especially in the dry season. There is a potential for large-scale cowpea production during the dry season, using the existing irrigation facilities, promoted specially to stimulate the production of wheat in the area and residual moisture in wetlands and river beds (Singh and Blade, 1997).In the early 1990s, IITA, in collaboration with the International Livestock Research Institute (ILRI), initiated a breeding program to develop dual-purpose cowpea varieties that produced both grain and fodder to suit the diverse needs of farmers in the semi-arid regions.

A recent breakthrough in cowpea breeding was the development of dry-season dual-purpose cowpea varieties adapted to the conditions of the semi-arid zone. Several dry-season cowpea varieties were evaluated in irrigated and wetland areas in on-farm dry season trials. These varieties have potential grain yields of over 1 t/ha with fodder yields of 4–10 t/ha when planted at the end of January to mid-February (Singh *et al.*, 1997) and harvested by the end of April to mid-May, when prices for cowpea grain and fodder are high. It is therefore, expected that this dry-season dual-purpose varieties would gain popularity among farmers in the semi-arid regions.

IMPORTANT VARIETIES

Crop improvement work undertaken at IGFRI and later under AICRP (FC) and at various SAU's resulted in development of high yielding, disease resistant varieties suitable for different cropping systems and agro-climatic conditions. A list of important varieties released since seventies are given below:

For Whole country : GFC-1; GFC-2; GFC-3 for kharif and GFC-4 for summer with 25-35 t/ha of fodder yield; Bundel lobia-1 (32-35 t/ha); UPC-287; UPC-5286 (30-45 t/ha)

For Haryana, Punjab and Delhi: FOS-1; FOS-10; K-395; K-585; IGFRI-S-450 (Kohinoor), C-88 (25-35 t/ha in Punjab) HFC-42-1 (Hara Lobia), HFC-128

For North India: UPC-5287 (30-45 t/ha); Russian Giant (30-35 t/ha)

For Uttar Pradesh and Karnataka: Russian Giant, IGFRI-S-978; IGFRI-S-985, DFC-1, C-152

For North, West & Central India: EC-4216 (35-40 t/ha) Bundel Lobia-2 (30-50 t/ha for North West) and UPC-4200 (34-45 t/ha for North East India) UPC-607, UPC-610,.UPC-612, UPC-616

For Gujarat: Charodi 14-20, Charodi 26-28

For Southern states and West Bengal: CO-1, Russian Giant, EC-4216, TNFC-9901

For Himachal Pradesh: PC-1, PC-3, PC-12, PC-14 and PC-16, showing resistance to collar rot

SOWING

Cowpea possesses epigeal type of germin-ation and no seed dormancy has been observed in cultivated cowpeas. As the cotyledons rise above ground, hardy, cloddy and crusty surface layers may hinder seedling emergence. The crop performs well even with moderate seedbed preparation that can be achieved by one ploughing followed by harrowing. In intercropping situations, the seedbed preparation of main cereal crop is sufficient for cowpea. With well-prepared land, crop encounters fewer weeds. During *kharif,* the crop is sown after receipt of first monsoon rains, while the harvest of previous crops decides *rabi* and summer sowing time. Cow pea needs a seed rate of 40 kg/ha (line-sown), 50 kg/ha (broadcast sown) and 25-30 kg/ha (seed production purpose) and a row spacing 30-40 cm is optimum. Seed rate for intercropping with cereal forages depends on row proportion and varies from 15 to 20 kg/ha. A sowing at a depth of 3-5 cm has been found optimum for seedling emergence.

MANURING

Cow pea needs little N fertilization as crop nodulates freely with native *Rhizobia* and seed inoculation with efficient cow pea strains of *Rhizobium* enable crop to meet most of its N requirement. However a starter N dose of 15-20 kg/ha is given to meet the crop requirement till N fixation starts. The efficient *Rhizobium* strains of cowpea (Kumar, 1990) include: Cowpea 109 (IARI, New Delhi); UAS IV-5 & TAL 169 (UAS, Bangalore) DC 6 (Navgaon, Rajasthan) and GMBS 1 (TNAU, Coimbatore). Cow peas are nodulated by *Rhizobium japanicum*, which also nodulates soybean, groundnut, lima bean and kudzu and the amounts of N_2 fixed/ha/yr range from 65-130 kg/ha (Alexander, 1977). A Nitrogen economy of 35 and 50 (kg/ha) due to mixed cropping of *Pennesetum pedicellatum* and maize / sorghum/bajra with cow pea with additional 15 and 5 kg/ha P_2O_5 fertilization was observed (Singh, 1992). It has been observed that with high N fertilization and soils rich in organic matter, nodulatioh and N fixation are poor. In acid soils, lime and molybdenum application enhances the nodulation. Phosphorous fertilization was found to promote root growth and nodulation in cow pea and the extent of response is dependent on soil P status. In most of the areas, application of 40 kg P_2O_5/ha has been observed to be optimum dose. Further treatment of seeds with phosphate solubulizing bacteria (PSB) and *Vesicular arbuscular mycorrhiza* (VAM) was found useful in P nutrition of the crop. Potassium fertilization is not recommended and followed for the crop, however, in soils deficient in K and where the crop is raised for seed purpose, response to K fertilization is observed. In general application of FYM@10 t/ha in last ploughing along with 20-40 kg/ha of P_2O_5 is sufficient for raising successful crop. The crop also responds to S fertilization.

IRRIGATION

Cowpea is a deep-rooted crop, which under favorable soil conditions extends up to 1.5-2.0 m (Hall, 1999) and thus, can extract moisture from deeper layers of soil. Further under drought conditions cowpea is known for maintaining high leaf and xylem water potential by complete closure of stomata. It is for this reason, it usually don't require irrigation in *kharif* and is drought tolerant. However, at monsoon failure, a life saving irrigation is advantageous. During summer, 3-4 light irrigations of 3-4 cm depth each in 15 days are essential. Irrigation at 50% ASM is sufficient.

WEED CONTROL

Cowpea, due to its quick growing nature possesses weed-suppressing ability. However, many weeds emerge at initial stages and compete with resources and reduce crop yields by 50-70% in both fodder and seed crop. The critical period for weed competition in cowpea lies between 10-40 days after emergence of crop (Acosta, 1991). Cultural practices of proper land preparation, use of weed free seeds and maintenance of optimum plant population reduce weed problem to a great extent. One hand weeding at 20-30 days after sowing and/ intercultivation almost nullifies the weeds impact on crop yield. Use of Trifluralin @ 0.5 kg a.i/ha (pre-plant incorporation) in sole cowpea and Alachlor @ 2 l/ha in maize -cowpea mixed cropping were found to provide effective weed control. Intercropping of cowpea with annual cereal forages was also found effective in reducing weed menace over the sole cereals

Witch weeds (*Striga gesnerioides* (Willd.) Vatke.), a parasite on cowpea is prevalent in Indian Sub-continent (Lane and Bailey, 1992) which reduces crop yields. Introduction of *Striga* resistance in cowpea cultivars (Agarwal, 1991) and use of Dicamba (foliar spray), Metalachlor (Pre-emergence) herbicides (Eplee *et al*, 1991) along with trap cropping were effective against witch weed.

CROPPING SYSTEMS

Cow pea is rarely grown as sole crop and more often is intercropped with arable annual and perennial cereal forages for enhanced fodder production and improved nutritive value of fodder (Table 1). The cow pea intercropping with sorghum/ bajra/maize is most prominent in South-Central-Deccan-Plateau zone with variable rainfall comprising Central and Eastern Maharashtra and Southern MP (Rai, 1989); North Western Semi-arid and Arid zone (Southern Punjab, Haryana, N-W UP, N Rajasthan SE Rajasthan and Gujarat) and Central Semi-arid Vindhyan zone (Bundel Khand and Bagel Khand regions) and is regarded as the main source for fodder supply to animals. The same intercropping system is also followed in humid North Western regions (low temperate wet regions) of Jammu and Kashmir and Himachal Pradesh during June-

October after wheat harvest. Further, in Himalayan foothill regions (*terai*) maize cow pea-oat fodder cropping system is extensively followed (Chatterji and Maiti, 1992). The most important sequential food cropping system-involving cowpea is maize-potato-wheat-cowpea (fodder) of North India and rice-rice-cowpea in Kerala.

In coconut gardens of Kerala, cowpea beside, guinea grass and *Stylosanthes guianensis* are grown. Further in foothill regions of West Bengal and Assam, dinanath grass (*Pennesetum pedicellatum*) + cowpea/ rice bean/ *Stylosanthes* intercropping system is popular. Cowpea is also intercropped with guinea grass and napier bajra hybrids in Western and South India, respectively. The performance of component crops in intercropping varies with cultivars and for intercropping with maize (African tall), cowpea variety UPC-612 was found best at IGFRI.

Table 1. Intensive fodder crop rotations with cowpea for different agro- climatic zones in India

ZONES	CROP ROTATION	GREEN FODDER YIELD (t ha^{-1} annum^{-1})
NORTHERN	• Hybrid napier intercropped with berseem	211.7
	• Hybrid napier + lucerne	176.0
	• Berseem + japan rape - jowar + cowpea	170.5
	• Maize + cow pea-maize + cow pea - turnip-oat	190.0
CENTRAL AND WESTERN	• Hybrid napier + cow pea-berseem + mustard	286.3
	• Maize + cow pea-M.P. chari-berseem + mustard	197.2
	• M.P. chari-turnip-oat	192.3
	• M.P. chari + cow pea-berseem + mustard - Jowar + cow pea	168.6
	• Maize + cow pea- maize + cowpea-oat-maize + cow pea	168.5

EASTERN	• Maize + cow pea-oat-bajra + cow pea	102.6
	• Jowar + cow pea-berseem + mustard-maize + cow pea	96.0
	• Maize + ricebean-berseem + mustard	111.5
	• Hybrid napier alone	144.2
SOUTHERN	• Sorghum + cow pea-maize + cow pea-maize + cow pea	110.7
	• Maize + cow pea-maize + cow pea-maize + cow pea	106.0
	• Guineagrass round the year	93.5

Soruce: Lal M. and Tripathi S.N. (1987)

Fodder production for intensive dairy farming

Overlapping system. The overlapping cropping system evolved by taking advantage of the growth periods of different species ensures a uniform supply of green fodder throughout the year. One such system continues for three years. The best rotation in this system is *berseem* + Japan *sarson* - hybrid napier + cowpea - hybrid napier; (October-April) - (April-June) - (June-October). Some of the important intensive fodder-crops - rotations and the expected yields from each are summarized in Table 3.

Fodder production in arable farming. There is ample scope for fitting in the short-duration fodder crops, either single or in mixture, with the other crops during the gap period between two main cash crops. Two distinct fallow periods are available for raising short-duration fodder crops, which provide adequate resources are available. In case of wheat-jowar rotation, gap periods between April and June and between October and November are available for each crop as fodder. Thus, in the first rotation. M.P. chari + cowpea, maize + cowpea, bajra + cowpea is successfully grown and an additional green-fodder yield to the tune of 300-350 q per ha is obtained. Similarly, in the second gap period (October-November), which is rather short, the growing of fodder turnips and short-duration mustard varieties help to get 250-300 q per ha of fodder. without disturbing the normal cropping systems (Table 2)

Table 2. Fodder crops crop rotations and the expected yields

1	Maize + cowpea - maize + cowpea + teosinte - *berseem* + mustard (300 q/ha) - (450 q/ha) - (1,000 q/ha)
2	Sweet sudan + cowpea - *berseem* + oats (1,000 q/ha) - (1,000 q/ha)
3	Hybrid napier + lucerne (1,250 q/ha) - (850 q/ha)
4	Maize + cowpea - *jowar* + cowpea - *berseem* + mustard (300 q/ha) - (400 q/ha) - (1,000q/ha)
3	Teosinte + bajra + cowpea - berseem + oats (1,000 q/ha) - (1,000 q/ha)
2	Sweet sudan + cowpea - mustard - oats + peas (1,000 q/ha) - (250 q/ha) - (500 q/ha)

Fodder production under dryland farming. In these areas, the farmers usually grow at least one crop in the *rabi* season for utilizing the residual moisture. Thus, there is great scope for raising two crops under such situations, firstly, the growing of a fodder crop which gets ready in 45-50 days after sowing (cowpea, guar, moth, etc.), yielding 150-250 q per ha of green fodder. After harvesting the fodder crops, crops such as gram, linseed, barley, wheat and safflower are raised on the conserved moisture.

Table 3. Stratified fodder-production potential of the best fodder-crops rotations in various regions (*All-India Co-ordinated Project for Research on Forage Crops*)

(A) Some promising rotations at various centers	Green fodder yield (q/ha)
• **Jhansi**	
1. Hybrid napier + cowpea - *berseem* + *sarson*	2,863
2. Maize + cowpea - M.P. chari - *berseem* + *sarson*	1,972
3. M.P. chari - turnips - oats	
• **Hyderabad**	
1. Hybrid napier + cowpea - Hybrid napier + cowpea - Hybrid napier + *berseem*	1,334
2. Maize + cowpea - *bajra* +cowpea + *berseem*	1,267

3. *Madikattujonna* + cowpea - jonna (ratoon) + cowpea - *berseem*	1,098
4. Hybrid Napier + *guar* - lucerne	2,529
5. Maize + cowpea - maize - cowpea - oats - maize + cowpea	1,685
• **Kalyani**	
1. Maize + cowpea - *P. pedicellatum* - oats	1,308
2. Maize + cowpea - rice bean - *berseem* + *sarson*	1,115
3. Maize + cowpea + *jowar* + *cowpea* - *oats*	884
• **Kanke**	
1. Maize + cowpea - oats - *bajra* + cowpea	1,026
2. *Jowar* + cowpea - *berseem* + *sarson* - maize + cowpea	960
3. *Bajra* + cowpea - *berseem* + *sarson* - maize + cowpea	959
• **Jorhat**	
1. Maize + cowpea - maize - *jowar* - oats	664
• **Hissar**	
2. *Berseem* + Japan rape - *jowar* + cowpea - *jowar* + cowpea	1,705
• **Coimbatore**	
1. Sorghum + cowpea - maize + cowpea - maize + cowpea	1,107
2. Maize + cowpea - maize + cowpea - maize + cowpea	1,060
• **Palampur**	
1. Maize + cowpea - lucerne + oats + *sarson*	844
2. Maize + cowpea - turnip - oats + pea - cowpea	833
3. M.P. chari + cowpea - oats + pea - cowpea	782
• **Jabalpur**	
1. Hybrid napier intercropped with cowpea - *berseem* and cowpea	1,761
2. M.P. chari - cowpea - *berseem* + *sarson* - *jowar* + cowpea	1,686

(B) Cowpea with cereal fodder crops (some of the important grass-legume mixtures)

Maize	***Jowar***	***Bajra***
Maize + cowpea (summer *kharif*)	*Jowar* + cowpea (summer *kharif*)	*Bajra* + cowpea (summer *kharif*)
Teosinte + cowpea		

Cultivated grasses

Hybrid napier	**Para**	**Guinea**
Hybrid napier + cowpea	Para + rice bean	Guinea + cowpea
Hybrid napier + maize + cowpea		
Setaria	Rhodes	Sudan grass
Setaria + cowpea	Rhodes + cowpea	Sudan + cowpea

Yield and fodder quality

Cowpea is a single cut fodder crop usually taken at 70-90 days of planting; however, summer crop can give 2 cuts. The crop is usually harvested along with the intercropped cereal, irrespective of its age. A green fodder yield of 30-45 t/ha of superior quality is obtained and is free from anti-nutritional factors. It makes good silage with additional molasses (30-40kg/t), usually ensiled with jowar and maize in 1: 2 ratio. Cowpea makes good hay and its seed is used as concentrate in animal feeds.

The crude protein content of cowpea fodder lies in the range of 16-21% and vary with cultivars, for instance, IGFRI-143 has 19%, whereas, IGFRI-450 has 21% (Rekib, 1992). The nutritive value of fresh biomass of cowpea (Dry matter basis) is 12.5%, digestible crude protein (DCP), 62.0% total digestible nutrients (TDN), 2.7 M cal/kg of digestible energy (DE) and 2.2 M cal/kg metabolizable energy (ME). The corresponding values for cowpea hay and seeds are 10.2, 53, 2.4 and 1.9 and 21.1, 84.0, 3.7 and 3.3, respectively (Bhattacharya and Sharma, 1992).

In many areas cowpea is the only available high quality legume hay for livestock feed. Digestibility and yield of certain cultivars have shown comparable to alfalfa. Cowpea may be used green or as dry fodder. It is also used as a green manure crop, a nitrogen fixing crop, or for erosion control. Similar to other grain legumes, cowpea contains trypsin inhibitors which limit protein utilization.

SEED PRODUCTION

Cowpea is self-pollinated crop. The suitable sites for seed production lie in North West and South West India that include the states of Andhra Pradesh, Tamil Nadu, West & Terai regions of Uttar Pradesh, Haryana, Punjab and Gujarat. The seed production potential of cowpea is 15-20 q/ha. Foliar spray of Phosphon-D 100 ppm, Planofix 50 ppm at 30 days after sowing (Dinesh Kumar *et al.*, 2000) and KNO_3 (2 kg/ha) at flower initiation stage (Sinha, 1980) were found to enhance seed yield.

The certification standards for seed crop are: minimum germination: 75%; minimum seed purity: 98%; maximum weed seeds: 0.1%; maximum other crop seeds: 0.05%; maximum inert matter: 2.0%; moisture (maximum), 9% in ordinary containers and 8% vapor proof containers.

REFERENCES

Acosta, M.1991. Determination of the critical period of competition of weeds in cowpea bean (*Vigna unguiculata* (Walp.)). *In: Alanje, Panama. Ciencia Agropecuaria Instituto de Investigation Agropecuaria de Panama* 7: 43-48

Agarwal, V.D. 1991. Research on cowpea striga resistance at IITA. *In:* Kim. S.K (Ed.). *Combating striga in Africa*, IITA, Ibadan, Nigeria, pp. 90-95.

Alexander.M.1977. *Introduction to Soil Microbiology* (Second Edition) New York; Wiley.

Bhattacharya.N.K. and Sharma, K. 1992. Forages for goat production. Pp. 237-253. *In: Proc. National Symposium on Strategies for forage production and Improvement by 2000 AD.* R.S. Paroda, S.K. Arora and A.S. Faroda (Eds.). 21-23 Sept. 1989. CA, Vellayani, Trivandrum.

Bressani, R. 1985. Nutritive value of cowpea. pp 353–360 *in* Cowpea: Research, Production and utilization, S.R. Singh and K.O. Rachie (Eds.). John Wiley & Sons, New York, USA.

Chatterjee, B.N. and Maiti. S. 1992. Cultivated fodders in existing cropping systems. pp. 125-134 *In: Proc. National Symposium on Strategies for forage production and Improvement by 2000 AD.* R.S. Paroda, S.K. Arora and A.S. Faroda (Eds.). 21-23 Sept.1989. CA, Vellayani, Trivandrum.

Delgado, C., Rosegrant, M., Steinfeld, H., Ehui, S. and Courbois, C. 1999. *Livestock to 2020: the next food revolution. Food, Agriculture and the Environment Discussion Paper 28.* Washington, DC: PRI/FAO/ILRI.

Dinesh Kumar, Seth, Rakesh and Padmavati.2000. Increasing seed production of fodder cowpea through efficient management. *Indian Farmg.* 50(3): 10-12.

Donal, M., Ball, D.M., Carl. S., Hoverland, J.S. and Lacefield, G.D. 1991. *In: Southern Forages.* Williams Printing Co., Atlanta, Georgia, and USA. pp: 52.

Eplee. R.E, Westbrooks, R.G. and Norris. R.S.1991. Chemical control of Striga. pp.61-68. *In:* Kim. S.K (Ed.). *Combating striga in Africa,* IITA, Ibadan, Nigeria.

Gallaher, R. N., and L. Hawf. 1997. Roles of conservation tillage in production of a wholesome food Supply. pp 23-27. In R.N. Gallaher and R. McSorley (eds.). Proceedings 20th Annual Southern Conservation Tillage Conference for Sustainable Agriculture. Special Series SS-AGR-60. University of Florida Cooperative Extension Service, Gainesville, Florida.

Grantz, D.A and Hall, A.E.1982. Earliness of an indeterminate crop, *Vigna unguiculata* (L.) Walp., as affected by drought, temperature and plant density. *Aust.J Agric Res.* 33: 531-540

Hall, A.E.1999. Cowpea.*In: Crop Yield: Physiology and processes*, pp: 355-373 D.L. Smith and C. Hamel (Eds.). Springer-Verlag Berlin Heidelberg.

Hazra, C.R. and Sinha, N.C. 1996. *Forage seed production. A Technological Development.* South Asian Publishers Pvt. Ltd. New Delhi-110 014.

Kandaswamy, P., Muthuswamy, P. and Krishnamoorthy, K.K. 1976. Effect of stage of cutting on the crude protein, phosphorus content and yield of cowpea (*Vigna sinensis* L.). Madras Agric. Jour. 63(3):178-180.

Kumar, V. 1990. Efficient Rhizobium strains for different pulse crops. *In: Advances in Pulse Production Technology*.pp.180. Jeswani, L.M. and Baldev, B. (Eds). ICAR, New Delhi.

Lal Menhi and Tripathi S.N. 1987. Strategies and technologies to maximise fodder production. *Indian Farmg*, (11): 16.

Lane, J.A and Bailey, J.A. 1992. Resistance of cowpea and cereals to parasitic angiosperm striga. *Euphytica*.63: 85-93.

Lal, R.B., D. Rajat, and D.J. Minson. 1978. Legume contribution to the fertilizer economy in legume-cereal rotations. *Indian Jour. of Agric. Sci.* 48(7):419-424.

Mortimore, M.J., Singh, B.B., Harris, F. and Blade, S.F. 1997. Cowpea in traditional cropping systems. *In advances in cowpea research.* Copublication of International Institute of tropical Agriculture (IITA) and Japan International Research Center for Agricultural Sciences (JIRCAS). Chapter 9. Sayce Publishing. Devon, UK. pp. 99-113.

Moss, E.V. and Hoffman.G.J.1977. Crop salt tolerance-Current assessment. *J.Irrig. Drainage*. Div. Am. Soc.Civil.Eng.103: 115-134

Rai, S.D. 1989. Intensive forage production in southern India. *Forage Production in India*, Singh, Punjab (ed.), Rangeland Management Society of India, IGFRI, Jhansi.pp.11-14.

Rekib, A. 1992. Forage quality with special reference to animal productivity. pp. 254-264. *In: Proc. National Symposium on Strategies for forage production and Improvement by 2000 AD.* R.S. Paroda, S.K.Arora and A.S.Faroda (Eds.).21-23 Sept.1989. CA, Vellayani, Trivandrum

Relwani, L.L., Kurar, C.K. and Bagga, R.K. 1970. Varietal trial on cowpea (*Vigna sinensis*) for fodder production. *Indian Jour. Agric.* 15(2):166-168.

Singh, B.B. 1993. Cowpea breeding: Archival report (1988-1992). Crop Improvement Division, IITA, Ibadan, Nigeria. Soil Survey Staff. 1984. Official series description of the Millhopper series. United States Government Printing Office, Wash. D.C.

Singh, B.B. and Blade, S.F. 1997. Potential of dry season cowpea in West Africa. pp 227–234. *In* Technology options for sustainable agriculture in sub-Saharan Africa, T.Bezuneh, A.M. Emechebe, J. Sedgo, and M. Ouechaogo (Eds.) OAU-STRC SAFGRAD, Ouagadougou, Burkina Faso.

Singh, B.B., Chamblis, O.L. and Sharma, B. 1997. Recent advances in cowpea breeding, pp. 30–49 *In:* Advances in Cowpea Research, (edited by B.B. Singh, D.R. Mohan Raj), 82: 79-242 K.E. Dashiell, and L.E.N. Jackai. IITA, and Japan International

Research Centre for Agricultural Sciences (JIRCAS) copublication. Available at IITA, Ibadan, Nigeria

Singh Panjab. 1992. Forage production through various farming systems. pp.119-124. *In: Proc. National Symposium on Strategies for forage production and Improvement by 2000 AD.* R.S Paroda, S.K. Arora and A.S. Faroda (Eds.). 21-23 Sept. 1989. CA, Vellayani, Trivandrum.

Tarawali, S.A., Singh, B.B., Peters, M. and Blade, B.F. 1997. Cowpea haulms as fodder. *In*: Agriculture (IITA) and Japan International Research Center for Agricultural Sciences (JIRCAS). Chapter 27. Sayce Publishing. Devon, UK. P 313-325. 177

Warrang. M.O.A and Hall A.E. 1984a. Repoductive responses of cowpea (*Vigna unguiculata* (L.) Walp.) to heat stress. I. Responses to soil and day air temperatures. *Field crops Res.* 8: 3-16.

CHAPTER - 13

COWPEA UTILIZATION IN INDIA

Pratibha Tewari, Sanju Choudhary and Nidhi Shekhawat

SUMMARY

Present paper reveals that food quality in terms of protein content and amino acid composition has become an important social issue in view of widespread malnutrition. Overcoming such malnutrition especially in third world countries is an important goal of every government in their respective regions. Raising food production and conserving the produce will help in eradicating malnutrition to a larger extent, simultaneously exploiting lesser known foods for human consumption hence reducing the cost of foods. Cowpea is one such food crop, which can help eradicate malnutrition among children and adults at a lower cost. Research efforts therefore, in cowpea production should now also be oriented towards better product development, storage and marketing. There is need to improve the post-production technology of cowpea so that its losses may be reduced to a minimum. Traditional milling technology of cowpea needs to be modernized. Almost 80.0% of the grain legumes produced is consumed as dhal/fried grains (whole/split/milled). There is need to develop suitable products from grain legumes and suitable packaging technology for processed cowpea.

INTRODUCTION

Cowpea also called as black eyed pea and china bean is believed to be native to India and the Middle East but in early times it was cultivated in China (Encyclopaedia Britannica, 1993). In India white, brown and black seeded varieties are cultivated; nutritionally they may however not differ. It is extensively cultivated in Indian belt for food, fodder, green manure and cover crop. Though cowpea is nutritionally at par to other pulses it is, however, considered as poor man's food.

CHEMICAL AND NUTRITIONAL COMPOSITION

The chemical composition of cowpea varies with variety, soil type, climate and the age. Like other pulses, it is also a rich source of protein in the diet; also rich in digestible carbohydrates. It is considered a nutritious source of protein content (23 to 25%), fat (1to1.3%), fibre (3.8%), carbohydrate (54%). As in most legumes, the amino acid profile of this crop complements the cereal grains.

Carbohydrate

Cowpea contains 54.5% carbohydrate, however some of the varieties in drier zones contain only 39% carbohydrate (APR, 2003), which suggests that in general protein content of drier zone cultivars is higher to cultivars of other zones. Almost 3.8g of this carbohydrate is present in the form of fibre cellulose which itself does not contribute to the nutritive value of food but helps in the mechanism of digestion and elimination of waste. The contraction of muscular walls of the digestive tract is stimulated by the fibre, thus counteracting the tendency to constipation.

Protein

Pulses are rich source of protein in a vegetarian diet but are deficient in methionine content. Cowpea is no exception to the limiting quality of pulse protein i.e. it is deficient in methionine but is rich in essential amino acids like, lysine. Hence, it can supplement proteins of cereals. The cultivars available in India contain 24.1% protein, while in drier zone, the protein content varies from 23.80 to 26.20% with a mean of 25%. The tender leaves of the plant are good source of protein (up to 25%) and their protein quality is high.

Fats and lipids

Pulses with their higher content of total fat (visible and invisible) contribute in combination with cereals to cater need of the essential fatty acid of an adult. The Indian cultivars had 1% fat whereas in drier zones, the fat content of pulses ranged from 1.75 to 4.05% with a mean value of 3.3%.

Vitamins and minerals

Cowpea contains 3.2g of minerals. It contributes significantly to B vitamin intake. Germinated cowpea is a fair source of vitamin C. Cowpea contains 77 mg calcium, 8.6 mg iron and 414 mg phosphorus per 100g grain.

NUTRITIVE VALUE

Cowpea is considered a poor man's pulse, though its nutritional value matches several other high esteemed pulses. A comparative account of cowpea in relation to certain other pulses is given in Table 1.

Table 1. Nutritive value of cowpea in relation to bengal gram and green gram seeds

NUTRIENTS	COWPEA	BENGAL GRAM	GREEN GRAM
Protein (g)	24.1	17.1	24
Arginine mg/g	420	570	500
Histidine mg/g	200	160	170
Lysine mg/g	430	440	460
Tryptophan mg/g	70	50	60
Phenylalanine mg/g	320	360	350
Tyrosine mg/g	230	180	100
Methionine mg/g	90	80	80
Cystine mg/g	80	80	60
Threonine mg/gm	230	220	200
Leucine mg/g	480	580	510
Iso leucine mg/gm	270	320	350
Valine mg/g	310	310	320
Fat (g)	1.0	5.3	1.3
Minerals (g)	3.2	3.0	3.5

Magnesium mg/100g of edible portion	210	119	127
Sodium mg/100g of edible portion	23.2	37.3	28
Potassium mg/100g of edible portion	1131	808	843
Copper mg/100g of edible portion	0.87	1.18	0.39
Manganese mg/100g of edible portion	1.34	1.21	2.47
Molybdenum mg/100g of edible portion	1.890	0.154	0.304
Zinc mg/100g of edible portion	4.6	6.1	3.0
Chromium mg/100g of edible portion	0.029	0.008	0.014
Sulphur mg/100gm of edible portion	165	179	188
Calcium mg/100g of edible portion	77	202	124
Iron mg/100g of edible portion	8.6	4.6	4.4
Phosphorous mg/100g of edible portion	414	312	326
Vitamins			
Carotene (μg)	12	189	94
Thiamine (mg)	0.51	0.30	0.47
Riboflavin (mg)	0.20	0.15	0.27
Niacin (mg)	1.3	2.9	2.1
Total B6 (mg)	-	-	-
Vitamin C (mg)	0	3	0
Folic acid			
Free (μg)	69.0	34.0	-
Total (μg)	133.0	186.0	-
Fibre (g)	3.8	3.9	4.1
Carbohydrate (g)	54.5	60.9	56.7
Energy (KCal)	323	360	334

All values are per 100 g of edible portion

ANTI NUTRITIONAL FACTORS

Legumes contain some anti nutritional factors (ANFs), the same substances although nontoxic can develop adverse physiological disorders and interfere with the utilization of nutrients. The ANFs are protease inhibitors, lecithins, goitrogens, anti vitamins, phytates, saponins, oestrogens, flatulence factors, allergens and lysinoalanins (Liener, 1981). Some other ANFs are cyanogens, favism factors, lathyrism factors, amylase inhibitors, tannins, aflatoxins and pressor amines. Although only a few legumes may contain all these ANFs, many contain a few of them. Some legumes contain trypsin inhibitors. Trypsin inhibitors reduce protein digestibility. Heating or autoclaving the legumes may help destroy the trypsin inhibitors and improve protein quality. Most of the ANFs are heat-liable hence, legumes are generally consumed after cooking, and it would not produce any major health hazard. Simple soaking and heating may not however, easily remove the heat stable compounds (polyphenols and phytates). The same could be reduced by germination and/or fermentation. In cowpea anti-nutritional problems can be easily avoided with appropriate dehulling and heat treatment. Legumes are rich source of polyphenolic compounds. Till recently, some of these (e.g. tannins) were considered as anti-nutrients due to their adverse effects on protein digestibility.

However, considerable interest in the antioxidant activity is generating now-a-days, and on their potential health benefits, especially in the prevention of cancer and cardiovascular disease (Menon, 2000). Dark colored legumes like red kidney beans, black beans, black gram and soybean have higher amount of these polyphenolic compounds.

COOKING QUALITY

Almost all foods consumed by human beings are subject to cooking, the exception being fruits and some green vegetables, which are used raw as salads and chutneys. Nutritional content of foods essentially relates to cooked food that reaches the table. Cooking practices vary from one region to another in the country and even from one household to another. Cooked food therefore, cannot be generalized and information on the effect of a wide range of cooking practices on nutritive value of food varies over the regions. Wet and dry methods of cooking are practiced in India. Wet method of cooking means boiling, steaming or pressure cooking; dry method of cooking is practiced at high temperatures like frying, roasting and baking.

Cooking of cowpea has both good and adverse effects. The nutrient losses during ordinary cooking are not of great concern, as the same causes little loss of protein, fat or carbohydrate. Some protein and minerals, especially Na, K and Ca may be lost if the food is cooked in water containing salt and the cooking water is discarded, it is, therefore, advisable to cook in either minimum amount of water or to use the cooking water in soups and gravies. Certain amount of minerals and vitamins are lost even during preliminary washing before cooking, hence, washing of cowpea with large quantity of water and repeated washing may be avoided. Washing may remove as much as 40% of thiamine and nicotinic acid. More than the minerals, there are vitamins especially those of water-soluble B group showing a greater loss during cooking. Vitamin A and carotene are not lost during wet cooking. However, cooking in acid media with tamarind and other acids has a protective effect against vitamins. Excessive heating may compromise nutritive value of proteins, particularly if the heating is done in presence of free reducing sugars since; in this process lysine combines with carbohydrate rendering lysine unavailable.

Cooking also has some beneficial effects as it improves the appearance and palatability of foods and confers new flavours. Cooking also destroys harmful food-borne microorganisms. In general, quality of proteins are improved after cooking since their access to enzymes and their digestibility are improved. Thus, cooking as a practice has several advantages in improving the quality, digestibility and palatability. In addition to it, cooking may also reduce the contents of vitamins and minerals however; some precautionary measures (cooking in limited water and in presence of acids) may help preserve vitamins to a greater extent. Study on effect of germination at 25° or 30°C for 24 hrs and the effect of processing steps (heated-air drying, decortication and cooking) revealed that germination had little effect on amino acid profile of cowpeas. *In vitro* protein digestibility was not improved significantly by germination or by decortication but was improved by cooking. Germination may significantly increase the contents of niacin, thiamin and riboflavin. Decortication resulted in upto 30% loss in niacin while thiamin content was reduced by 41% on cooking (Nnanaa and Phillips, 1989). The use of cowpea in cereal-based food may improve their nutritional quality. Study revealed that cowpea could be used in making vermicelli up to 25% without altering their physicochemical and sensory characteristics (Coulibaly *et al.*, 2003).

COWPEA PRODUCTS

Pulses enjoy the distinction of being protein-rich food and form an important constituent of diet. Bengal gram, black gram, cowpea, green gram, lentil, red gram and dried peas are some pulses used for various preparations. Pulses form excellent supplement to cereal diet, as they are rich source of lysine. Thus, the combination of cereals and pulses in the diet improves the protein quality. Legumes are consumed in various ways as greens (green, roasted, boiled, fried, crushed and cooked); as mature dried grain (boiled, boiled and fried, cooked as dhal, germinated). De-husked splits (cotyledons) as dhal/sambhar or converted into flour and used in a number of preparations.

Process involved in preparation and use of cowpea are soaking, cooking, roasting, puffing, extrusion cooking, germination, fermentation, canning, expression and expelling. The utilization pattern of cowpea can be grouped as follows:

- Whole/green cowpea (fresh, boiled and roasted);
- sprouted and germination (boiled and fried);
- puffed and roasted (spiced/salted);
- milled and cooked (steamed, boiled and fried); and
- fermented products (idli, dhokla etc).

A number of cowpea preparations based on common Indian diets are possible to be prepared. (Goyal, 1996). A resume of nutritional value of these products, cooking time, cooking methods, cost of the product and the keeping time are given in Table 2.

Table 2. Cowpea Products Based on Indian Diets

Preparation	Ingredients	Cooking Method	Nutritional quality	Cost of the product (Rs/serving)	Cooking Time (min)	Keeping Time (hr)
Whole/Green Cowpea						
Cowpea vegetable	Cowpea, onion, garlic, ginger, green chilli, tomato, coriander leaves, salt and spices and fat.	Soaking and pressure-cooking.	Protein-7.37 Fat-15.4 CHO-21.72 Energy-255.25	3.00	30*	8 hrs
Cutlets	Cowpeas, boiled potato, onion, green chilli, ginger, coriander leaves, cumin (roasted) and salt.	Soaking and frying	Protein-31.7 Fat-26.45 CHO-136.2 Energy-908.5	8.00	30*	8 hrs
Bonda	Potato, cowpea, coriander leaves, ginger, gram flour, green chilli, fat, salt and spices.	Boiling and frying	Protein-24.05 Fat-53.4 CHO-79.75 Energy-894.5	5.00	30	8 hrs
Cowpea paneer	Cowpea, paneer, tomato, onion, garlic, green chilli, ginger, coriander leaves, fat, cumin seeds, salt and spices.	Boiling, grinding and frying	Protein-10.58 Fat-19.53 CHO-18.81 Energy-293.5	5.00	30	6-7 hrs

Cowpea pods vegetable	Green cowpea, onion, tomato, oil	Roasting and seasoning	Protein-4.6 Fat-5.3 CHO-15.3 Energy-127.5	3.00	20	5-6 hrs
Sprouted and Germinated						
Sprouted pulses salad	Green gram, bengal gram, cowpea, groundnut, salt, green chilli, onion and lemon juice.	Soaking and sprouting	Protein-47.05 Fat-23.95 CHO-111.7 Energy-851	7.00	10*	5-6 hrs
Milled and cooked products						
Wheat and Cowpea vadai	Whole wheat dalia, cowpea dalia, potato, ginger, green chilli, onion, spices and oil.	Soaking, grinding, and frying.	Protein-5.73 Fat-25.37 CHO-30.16 Energy-371.87	3.00	30*	8 hrs
Kachori	For crust-Maida, fat and salt. For filling-cowpea, spices, salt, coriander leaves, green chilli and oil.	Kneading, boiling and frying	Protein-8.12 Fat-25.47 CHO-33.07 Energy-394	2.50	45	8 hrs
Porridge	Wheat dalia, cowpea dalia, milk and sugar	Soaking and cooking	Protein-8.48 Fat-5.42 CHO-31.15 Energy-207.1	2.50	150*	4-5 hrs

Laddoos	Semolina, cowpea flour, sugar, khoa, grated coconut, cardamom and ghee.	Roasting, and mixing	Protein-14.64 Fat-21.26 CHO-76.09 Energy-554.6	5.50	45	8-10 hrs
Biscuits	Maida, cowpea flour, ghee, powdered sugar, milk, baking powder.	Sifting, mixing and baking	Protein-6.06 Fat-21.3 CHO-35.26 Energy-357.1	4.00	60	8-10 days
Puri	Wheat flour, cowpea flour, oil and salt.	Kneading and frying	Protein-9.04 Fat-23.17 CHO-30.97 Energy-368.5	2.28	60	10-12 hrs
Batti	Wheat flour, cowpea flour, ghee and salt.	Kneading and baking	Protein-9.04 Fat-30.67 CHO-30.97 Energy-436	5.00	30	5-6 hrs
Stuffed parantha	Wheat flour, cowpea flour, potato, oil, green chilli, coriander leaves, turmeric powder and salt.	Kneading, stuffing and shallow frying	Protein-9.52 Fat-5.7 CHO-37.75 Energy-240.1	1.50	30	5-6 hrs
Mathari	Maida, cowpea flour, oil and salt.	Kneading and frying	Protein-8.77 Fat-27.97 CHO-32.09 Energy-415.2	3.00	20	10 days

Paustick kichari	Rice, green gram dhal, cowpea dalia, oil, onion, carrot, peas, spinach, turmeric powder and salt.	Soaking and cooking	Protein-6.29 Fat-15.39 CHO-33.51 Energy-298.4	3.00	20*	5-6 hrs
Muffin	Maize flour, cowpea flour, sugar, baking powder, salt, milk, egg, ghee and cardamom powder.	Beating and baking.	Protein-12.79 Fat-11.27 CHO-55.55 Energy-374.6	3.00	60	2-3 days
Rabri	Butter milk, bajra flour, cowpea flour, salt, cumin.	Boiling	Protein-4.77 Fat-2.25 CHO-12.95 Energy-90.9	1.00	30	5-6 hrs
Chikki	Cowpea flour, sugar, khoa, ghee and milk	Frying and roasting	Protein-17.21 Fat-37.17 CHO-73.45 Energy-697.3	9.00	90	12 hrs
Dahi vada	Green gram, black gram, cowpea, curd, salt and oil	Soaking, grinding and frying	Protein-8.81 Fat-12.36 CHO-18.9 Energy-221.8	2.50	60*	1-2 days
Halwa	Cowpea, ghee, sugar and cardamom powder	Soaking, grinding and frying	Protein-7.23 Fat-30.3 CHO-46.17 Energy-486.3	6.00	120*	5-6 hrs

Moong dal vadi	Green gram dhal (dehusked), cowpea, salt, green chilli and oil.	Soaking, grinding, blending and sun drying	Protein-80.42 Fat-8.87 CHO-187.39 Energy-1149.8		190*	5-6 months
Papad	Green gram dhal flour, cowpea flour, baking soda, salt, black pepper and cumin seeds.	Kneading, rolling and drying.	Protein-277.72 Fat-19.25 CHO-662.95 Energy-3931.5	-	240*	8-9 months
Gattaa curry	Gram flour, cowpea flour, oil, cumin powder, omum seeds, salt and red chilli powder.	Kneading, boiling and frying	Protein-11.22 Fat-16.65 CHO-28.57 Energy-308.7	2.50	30	5-6 hrs
Poshtik soup	Cowpea, green gram dhal, spinach, salt, cumin seed, pepper, butter.	Pressure cooking, grinding and boiling.	Protein-12.64 Fat-12.87 CHO-29.31 Energy-283.6	3.50	30	2-3 hrs
Fermented products						
Dhokala	Cowpea flour, onion, ginger, coriander leaves, salt, oil, cumin seeds, sodium bicarbonate and curd.	Steaming	Protein-15.63 Fat-17.69 CHO-42.58 Energy-392.2	5.00	60	5-6 hrs
Idli	Rice, black gram dhal, cowpea and salt.	Soaking, grinding, fermenting and pressure-cooking.	Protein-6.17 Fat-0.34 CHO-27.05 Energy-136	1.00	30*	5-6 hrs

Puffed and roasted products						
Chutney	Fresh coconut, roasted cowpea, green chilli, salt, curd, mustard seeds, curry leaves and oil.	Roasting, Grinding and tempering	Protein-4.3 Fat-13.5 CHO-9.45 Energy-176.3	2.00	10	2-3 days
Burfi	Cowpeas, khoa, sugar and coconut powder	Soaking, roasting, grinding and setting	Protein-13.4 Fat-7.0 CHO-54.45 Energy-334.6	5.00	30*	2-3 days
Poshak	Whole wheat, cowpea and groundnut.	Roasting and grinding	Protein-9.04 Fat-3.6 CHO-31.48 Energy-194.7	1.00	60	7-10 days
Poshak laddoo	Poshak mixture, powdered sugar, ghee and cardamom powder.	Roasting	Protein-9.04 Fat-13.6 CHO-61.3 Energy-404.1	3.00	20	6-8 days

Units of nutritional quality: Protein, g; Fat, g; CHO, g; Energy, Kcal.

* Excluding soaking time (2-3 h- overnight).

SUPPLEMENTARY FEEDING TO CHILDREN

The infant and the pre-school children are extremely vulnerable group. The rate of growth and development of infants and pre-school children largely depend on extent and adequacy of the diet consumed by them. A majority of infants and pre-school children in the developing countries consume inadequate diets. Consequently they grow at a sub normal rate and are stunted. They also suffer from signs and symptoms of dietary deficiencies. (Dean, 1953 and Jelliffe, 1966).

In view of the shortage in the supply of milk and other protein rich foods of animal origin in the developing countries, successful attempts have been made on the development of processed protein foods based on locally available legumes and their use as supplements to the diets of children. Recent investigations indicate that the utilization of oilseeds and legumes as supplementary feeds to the children could help meet the nutritional requirements and promote satisfactory growth of infants.

The nutritional value of cowpea, which is mainly the high protein level, rich in lysine, represents its major advantage for use in supplementary food for children besides, its easy availability and lower cost. A number of processed protein foods based on oilseeds, oilseed meal and legumes have been developed and studied. Weaning food from cowpea was formulated in Ghana. Evaluation of this product was conducted and it confirmed excellent contribution of the cowpea fortified weaning food to the health and nutritional status of the children (Phillips, 2004). At CFTRI, Mysore supplementary food named as Indian Multipurpose Food (MPF) was developed which is a blend of low fat groundnut flour and bengal gram flour (75:25) fortified with Vitamin A and D, thiamine, riboflavin and calcium carbonate. A Bal – Ahar named supplementary food, which is a blend of whole-wheat flour, groundnut flour and roasted bengal gram (70:20:10) fortified with calcium salts and vitamin has also been developed. Studies carried out with pre-school children have shown that supplementation of diet with MPF and Bal-Ahar brought about highly significant improvement in growth and nutritional status therefore helping to make up deficiencies in the diets of pre-school children. Another supplementary food based on a blend of wheat flour, green gram flour, groundnut and sugar/jaggery (30:20:8:20) has been developed at NIN, Hyderabad. A daily supplementation of 80g of this product has been found to bring about significant improvement in the growth rate of pre-school children. It is possible to effect further improvements in the diets by incorporation of 30g of pulses which will help to make up the deficiencies of proteins. These supplements are within the reach of the low-income groups of the population. When spontaneous lactic acid fermentation is applied with cowpea fortification to maize-based weaning foods, it is expected to further improve the nutritional quality, functionality and provide an alternative high protein weaning food for infant nutrition.

Cowpea fortification can be used to produce high protein weaning foods with improved product characteristics and its use as supplementary food will be further studied in AICRP on arid legumes.

PROCESSING AND STORAGE

Pulses are the major source of dietary protein in India. As of now, the annual production of pulses in India is about 15 million mt. India is one of the major pulse growing countries of the world sharing about 26 percent of the world pulses production. Pulses are under severe pressure from other remunerative crops. Resistance to subsequent attack by storage insect/pests like bruchids has suggested that steaming might be useful for preserving cowpea seeds stored for food. Studies on physical and functional properties revealed that seed viability was completely lost following exposure to steam; there were also significant decrease in the water absorption capacity and protein solubility of the treated seeds. A pronounced increase in cooked bean hardness was observed after the steaming process. In general, increasing drying temperature resulted in decreased cooked bean hardness.

Storage of grain legumes is necessary to meet the regular demand throughout the year. Safe storage therefore, helps to maintain continuity in supply. Although grain legume production is seasonal, however, processing/milling continues, grain legumes in commercial quantities are stored in modern storage structure i.e. flat godowns and in silos so that the loss during storage is minimized. Domestic level grain legume storage containers are traditional. The traditional containers are improved and replaced by modern airtight metal bins that are easily available in a variety of sizes. The metal bins can also be easily fumigated. About 80% of the storage loss is due to insects, rodents and micro-organisms. Grain legumes are specifically attacked by bruchids at the time of maturity itself. In high humidity and temperature (rainy season) conditions bruchids are difficult to be controlled. Farmers cannot afford to store grains even for seed purposes. The loss caused by bruchids may range from 10 to 30% (Singh and Verma, 1995). Most of the storage losses take place at farmer's level. Studies are conducted to control insect infestation of pulses using biogas fumigation in metal bins.

Controlled Atmosphere Storage

Different atmospheric conditions viz., CO_2 level of 60, 70 and 80 %, O_2 level of 4, 7 and 10 % and N_2 level of 16, 23 and 30 % were evaluated at 0.4, 0.6 and 0.8 kg/cm2 pressure level and air flow rate of 58 ml/min to control pulse beetle and rice weevil in green gram. The exposure time was 24, 36 and 48 hrs. The relative humidity (RH) and temperature were 71% and 30 ± 2°C, respectively. Results revealed that mortality occurred at higher CO_2 level and lower levels of O_2.

Thermal Treatment

Exposing grain legumes to 70-80°C temperatures for short duration can control insect infestation. The insects are killed at higher temperature.

Ventilation

Commercial storage structures should have arrangement for forced ventilation in order to prevent moisture accumulation and development of heat spots in the storage structures.

A study on farm trials showed that Hermetic storage in plastic buckets was the most effective method though too expensive. Solarisation and solarisation plus admixture of peanut butter were another method, which was tested by the farmers and found effective methods for storage.

POST-HARVEST TECHNOLOGY OF LEGUMES

Different operations involved in primary processing in legumes are cleaning, grading, drying/conditioning, storage and milling. Each operation has a definite purpose to achieve. These operations are common to most of the legumes and are briefly described.

Cleaning

Freshly threshed cowpea grain contain foreign materials like straws, twigs, shells, stones, dusts, immature grains and weed seeds. They are cleaned with rotary screen or vibratory screen cleaner-cum-grader. Lighter impurities are blown off with the help of blower/air aspirator. Large-size impurities are removed with scalper and small size impurities pass through the sieve and are removed.

Milling

Milling is a general term and refers to the reduction of grains into meal or flour. Milling is an overall process and it includes reduction in size, hulling, scarification, polishing, sorting, mixing and in some instances, also refers to certain chemical reactions. The yield of split pulses in traditional mills is only 65 to 70% in comparison to 82 to 85% potential yield. De-husking is a process that reduces the fiber content and improves appearance, texture, cooking quality, palatability and digestibility of grains. Pre-milling treatment plays an important role in improving *dhal* recovery. The proportion of seed coat, cotyledon and embryo in relation to whole seed in cowpea is 10.64%, 87.23%, and 2.13%, respectively (Salunkhe *et al.*, 1985).

Pre-milling Treatment

Pre-milling treatments are given to affect the gums present in between seed coats and cotyledon in order to: (a) lessen the husk; (b) ease of milling; (c) reduce breakage; and (d) improving the quality of split. The following four methods are used in pre-milling:

Dry Method: The dry method includes application of oil and water. In this method cleaned and graded grains are subjected to pitting to increase soaking. Pitted grains are mixed thoroughly with 1% oil and spread for sun drying in thin layer for 2-3 days. At the end of drying 2.5% water is sprayed and mixed thoroughly. For tempering, grains are heaped overnight. The grains are milled with roller machine.

Wet Method: In wet method, red earth slurry is applied on pulses. The grains are thoroughly mixed with a paste of red earth after soaking in water for about 12 hours and heaped for about 16 hours. The grains are spread in thin layers in drying yards for 2-4 days. Dried red earth is thereafter, removed by sieving. The grains are milled on power operated stone or emery-coated vertical *chakki* (manual mill). This method resulted in 95% de-husked and 75% *dhal* yield. Wet milling process is popular in South India. This method requires 5-7 days to complete processing for a batch of 6-7.5 mt.

Thermal Treatment: The cleaned grains are conditioned in two passes in a dryer (Louisiana State University [LSU] type) using hot air at about 120°C for a certain period of time. After each pass, the hot pulses are tempered in the tempering bins for about six hrs. Such hot air preconditioning of grains helps in loosening the husk significantly.

Chemical Treatment: The use of chemicals has demonstrated the potential to replace oil treatment to achieve at least the same and in some cases more recovery than traditional methods. Sodium bicarbonate and sodium chloride were used to treat pigeon pea grains. The grains soaked in 4-6% sodium bicarbonate solution for 0.5-1.0 hour and dried at 65°C to 10-15% moisture content gave 95% hulling efficiency and *dhal* yield was 80%. Possibility of using enzymes for treatment was also studied since some enzymes may break hemicelluloses, pentoses, hexoses etc. present in germ and seed coat, thus enhancing the ease of husk removal. The process is yet to be developed as complete package. Studies have shown that acetic acid (vinegar) was found effective in loosening the husk of pigeon pea.

Polishing: Polish is given to the de-husked and split grain pulses by treating them with a small quantity of oil and/or water. However, the products prepared from unpolished cowpea were found more acceptable than polished grain.

Packaging: Grains after cleaning, grading and milling are packed in polyethylene bags of different capacities for sale in the market. The polyethylene bags are sealed to provide modified atmosphere conditions to the product.

Milling Performance: Grain parameters such as size, shape and moisture content play an important role in deciding the milling performance and quality of milled grains. Milling operations yield *dhal* (50-80%), grits (5-20%), husk (10-25%) and powder (7-20%) depending on the legume crop and method of milling. Milling performance also depends on other factors viz. pre-milling treatment, energy inputs, handling, storage and machinery. Among them, the type of machinery helps to reduce the losses, decides the conveniences and improves the profit of the user. Techniques for de-husking and splitting are well known for centuries. However, it is important to identify the appropriate system to obtain maximum yield of the product in desired form with ease of operation, optimum energy costs and less drudgery (Kulkarni, 1993). The maximum recovery of dhal using modern and maximum possible recovery method of splits and average dhal (splits) yield for Cowpea is 89.5% and 84%, respectively (Kurien, 1977).

REFERENCES

Annual Progress Report 2003. All India Coordinated Research Project on Arid Legumes Project Coordinator's Unit, CAZRI, Jodhpur, India. pp: 1-177.

Coulibaly, S.S., Drame, D,, Traore, S., Diarra, M. and Bore, F.G. 2003. *Improvement of the Nutritional Quality and Sensory Evaluation Wheat/Cowpea and Millet/Cowpea Vermicelli.* Paper presented in 2nd *international workshop on food based approaches for a healthy nutrition in West Africa: the role of food technologists and nutritionists* held at Onagadongou in Burkina Faso. IRD, Wagenigen Univ, FAO.

Dean, R.F.A. 1953. *Plant proteins in child feeding.* In: Spl. Rep. Ser. No. 279. Medical Research Council, England.

Encyclopaedia Britannica 1993. The new encyclopedia Britannica, Enyclopaedia Britannica, Inc., USA. Vol 3. Pp: 697-698.

Goyal, M., 1996. Laboratory Manual of Basic Foods. College of Home Science, Bikaner, Rajasthan Agricultural University, Bikaner. Pp: 12-65, 86-102.

Jellife, D.B. 1966. *Infant nutrition in the sub tropics and tropics.* In: W.H.O Monograph No. 29(2) Ed. W.H.O, Geneva.

Kulkarni, S.D. 1993. *Development of food legumes machinery in India.* National Industry in India – Its Future Needs. National productivity council (NPC). Pp: 118-138.

Kurien, PP., 1997. "Grain Legume Milling Technology", Paper presented to *FAO Expert consultation on grain legume processing,* CFTRI, Mysore, India.

Liener, I.E., 1981. *Factors Affecting the Nutritional Quality of Soy Products. J. Amer. Oil chem. Soc.,* 58 : 406-415.

Menon, Krishnakumari, 2000. Legumes in Human Nutrition, 24(2), pp: 21-32.

Nnanna, I.A. and Phillips, R.D. 1989. Amino acid composition, protein quality and water-soluble vitamin content of germinated cowpeas (*Vigna unguiculata*). *Plant Foods Hum Nutr.* 39(2) : 187-200

Phillips, D. 2004. Improved Processing Technologies to Increase Cowpea Utilization and Improve Nutrition. Project, File:// A: / Cowpea CRSP.htm

Salunkhe, D.K., Kadam, S.S. and Chavan, J.K. 1985. Processing *and utilization.* In: Post harvest Biotechnology of Food Legumes. CRC Press, Inc. Boca Raton, Florida, USA. pp: 109-140.

Singh, J. and Verma, P. 1995. *Status of pulse milling in India.* Central Institute of post harvest engineering and technology (CIPHET), Ludhiana. pp: 22.

CHAPTER – 14

VEGETABLE COWPEA - FUTURE PROSPECTS

Dr.N.B.Gaddagimath

Plant protein is an important source of protein and an important component of nutrition especially in a developing county like India. Cowpea is an important pulse crop, which can add value to this need and a new hope in Nutritional Security globally.

Among all pulses, Cowpea, ***Vigna unguiculata*** L, is most productive, genetically versatile and phenotypically hard enough to sustain environmental shocks and stress.

Cowpea is considered as poorman's pulse as it is not liked by majority of people because of high flatulence factor, bitter content in grains and low profile flavor components. But it is a very potential crop with good productivity and adaptability.

Future of Cowpea turning out to be a most productive, competitive, nutritionally rich vegetable is too bright. It is recognized fact that vegetables are the major providers of fibre, green protein and biological water to the human diet contributing significantly to the nutritional security. Global awareness is increasing with regard to quality food and balanced diet especially in those countries where cereal consumption is more. The consumption of vegetable in every meal is increasing thereby decreasing the pressure on production and consumption of cereals where every meal is a cereal only, especially in developing countries.

With the growth of megacities, expecting urban population equal to or exceeding the rural population, the demand of vegetables is increasing in manifolds. The periurban horticulture is gaining importance for the benefit of urban consumers of large cities world over.

Vegetables are usually higher in productivity than other crops. These crops can provide more food per unit area, per unit time and unit labour and monies spent on production activity. They are highly remunerative. Most vegetable on an average can yield more than 10 M.T/Hec. within a period of 3 to 4 months.

Vegetables are the major crops affordable and sustainable source of nutrition for the masses, besides they are generally good sources of fibre, carotenoids, vitamin-c, foliate, potassium and other minerals, vitamins, minerals and bioactive compounds.

In view of these facts, the scope and prospects of genetically potential Cowpea crop turning out to be a promising vegetable pulse crop are too bright. Cowpea, *Vigna unguculata* (L.) Walp is a protein rich crop, demands more attention especially in a county like India, where majority of the people are vegetarians and suffer from protein malnutrition.

Scope

Very little attempt has been made to identify ideal cowpea pod for vegetable purpose. It is time that this crop should be thoroughly reviewed and the outlook of this crop should be changed into a most productive consumer acceptable vegetable crop. The market demands are also changing and there are ample opportunities for this crop to develop into a most popular vegetable crop made available throughout the year.

The variability available in this species and sub species is very potential and ample number of new lines, hybrids could be developed. The growth behavior, earlyness, adaptability of these species would make this crop popular among vegetables.

Sarpan Hybrid Vegetable Cowpea SCV 201

Sarpan Hybrid Vegetable Cowpea SCV 202

Sarpan Hybrid Vegetable Cowpea SC 13 (Dual Purpose Cowpea)

This crop needs to be attended scientifically and commercially to meet out market needs and demands.

Cowpea will turn out to be a parallelly popular bean vegetable like French bean, since cowpea has all quality and productive features of French beans. It can be popularized even in arid and semi arid regions, scanty rainfall areas where vegetable cultivation is in limited areas and under rain depended cultivation. This crop will turnout to be a most promising vegetable in these areas.

Genetic Background

The specie *Vigna unguiculata* L. has three cultivated subspecies, *unguiculata, cylindrica* and *sesquipedalis*. The subspecie *Vigna urguiculata* commonly cultivated for dry seeds, vegetable pods and grains. Plants are viny, sometimes erect, 20-30 cm long pods with small seeds. Subspecie *Cylindrica*, the plants are more erect, pods small 7-13 cms long and seeds small kidney shaped. Commonly grown Asperagus beans belong to this group and are mainly grown for vegetable purpose. Majority of cultivors of Sesquipedalis are vegetable types with long to extra long tender fruits. Fruits have low fiber, high protein. Seeds are awful and are not attractive. Genetic variability is enough in these sub species and need to be exploited for vegetable purpose. Crossability among *unguiculata* and *sasquipedalis* is very good and also with *cylindrica*. Identifying selective genes, characters among these species and bringing them under one genetic backup would be of great advantage for vegetable cowpea breeding. Scope is abundant and systematic approach is needed.

Asperagus beans the commonly grown vegetable cowpea is a vine and has many limitations in commercial cultivation. Though it has features of vegetable cowpea-having tenderlong pods, this cowpea is prone for lodging. This cowpea is grown mainly in rabi and summer and grain are not attractive and are limited to vegetable purpose only.

Strategies

Cowpea crop as such needs to be thoroughly reviewed and definite groups and objectives suitable to present and future market needs to be framed. Improvement programmes should be planned accordingly to meet out domestic and overseas market.

Development of consumer acceptable vegetable cowpea is very important. Majority of cowpea grown are not suitable for vegetable purpose. The tenderpods have bitter content and are not suitable for vegetable. The colour of pods is also important light green to white tender pods, stringless, very less fiber on pod sutures, easily dressable are accepted for vegetable. The pods should have good flavor component and good cooking quality. Even on steaming it should not turn out into paste. Semi hard, firm features on cocking needs to be retained. It should match french beans in all respect. Medium long and small pods should range between 6" to 9" in length and should maintain tender ness preferably for a longer duration.

Long pod types belonging to sesquipodalis subspecies need to be changed in their habit. They should be converted into bushy, compact and erect type with early bearing nature. This would commercially make these breeds more popular. The pods should be low fiber on pod sutures, grains should be soft and tender.

Second category of vegetable types should be of green fresh grain types, which are easily shellable. These kinds should have medium long pods of 8" to 10" with 8-12 bold seeds per pod. These seeds should be sweet with high glucocides, and do not taste bitter on fresh consumption. This kind of material is being consumed and should be made more popular for chats, fresh fried grains, semiboiled bold cowpeas with good flavor. Fast food market will have more scope for such products. It should be parallelly compared with groundnut grains.

Other kind with masaic grains, mixed colour, white bold types, white bold with black eye, brown eye and golden eye need to be developed. Scope for such kinds are bright. Today's fast food market is growing in manifolds. This market should be built up in these lines of new material. Demand is already there but material availability is lacking.

The third approach and strategy should be in developing dual-purpose cowpea. Viz, breeds suitable for vegetable as well as for grain purpose. This kind of material will have higher market value as it will go well as popular vegetable and if vegetable market saturation is there and prices fall, it's cultivation be continued for grain purpose. The vegetable types grown today are all not suitable for grains purpose. As the grains are flat, awfulshapes, colour-dull and do not have good cooking qualities. The new breed should have vegetable as well as grain qualities.

Nutritional values in vegetable crop are equally very important. Today the plant protein content reported in vegetable cowpea ranges from 3.5 to 4.5%. This needs to be seriously attended and improvement in protein contents is the need of the hour. Protein nutritional security would be the greatest contribution, this crop can make to mankind if consumer acceptable, vegetable breeds are developed and released to the market. Fresh grain vegetable types would add significantly to the high protein type of breeds, which will add value to the cowpea crop.

Photo insensitive, compact, erect breeds need to be developed to popularize vegetable types in future.

Salient features of Vegetable Cowpea

- Slender, green. Or white extralong, stringless with good cooking quality.
- Medium to small size, white or light green pods with medium sized seeds with good colour and flavour.
- Low fiber on pod sutures with low bitter content.
- Bushy compact or semicreeping types for all seasons.
- Bold sized grains for dual purpose for fresh grain vegetable as well as for grains.
- Short tender stringless pods for processing.
- Photoinsensitive for all season cropping.
- Breeds for wider adaptability.
- Pods with higher protein content.
- Pods with good flavor, taste and low fiber, low bitter content and high glycoside and proteins.

Breeding

Cowpea is basically a selfpollinated crop. Pureline Selection and plant to progeny selection in selective crosses and early generations have resulted in popular grain varieties. However, work in respect of vegetable cowpea is limited. The genetic variability reported in all these three sub species is significantly very high. Crossability within these subspecies is also high. Heterosis recorded in earlier studies is also significantly high and can be proposed for exploitation. But the crop nature doesn't support the commercial exploitation. However, new approaches are tried to exploit the heterosis in cowpea.

Conservation of heterosis would give a good scope in cowpea breeding Experiments are tried at Sarpan Agri-horticultural Research Center, Dharwad to produce sibling hybrid populations retaining significantly high level of heterosis in early segregating generations. These sibling hybrids can be reproduced every season satisfactorily using this new hybrid techniques of conservation of heterosis in earlier generation. Couple of hybrids are released for cultivation, which are performing very well and are accepted commercially. Sister lines with good general combining ability are isolated and are maintained. These groups differ a little within themselves and still produce high level of heterosis on crossing. The heterosis observed was mainly due to additive effect and the inbreeding depression is also low. This is taken advantage and few hybrids SC-13, SC-17, SCV-201, SCV-202, SCV-30 are released. Since the sisterlines do not vary much in grain features, the conserved heterosis in earlier filial generations is taken advantage. The experience of varietal genetic degradation in self pollinated crops and NSP's-recommendation to replace variety once in every three years to maintain yield stability has convincingly encouraged this new concept and is commercially viable. Yield levels are high, crop is early, vigorous and sustain environmental stress too. Sisterlines SCL-100, SCL-200, SCL-300, SCL-400, SCL-500, SCV-100, SCV-200, SCV-300 series are developed and maintained. New hybrids are under offing with unique vegetable and grain features. This concept would stabilize cowpea production and would further give new avenue for promotion.

Ideal Plant type

Compact, erect, good branching with 2-3 pods per spike, early, medium to long tender stringless white or light green pods with bold seeds or fleshy firm fruits, low fiber, low bitter content, high protein with high glycosides suitable for all season are the highly suitable plant type for vegetable cowpea.

Meanvalues of Fruit features of few Hybrid vegetable Cowpea breeds in offing 2003-04

Treat No.	Average Tender Fruit Length in cms	Average Mature fruit Length in cms	Average Tender Fruit weight in gms	Average Mature fruit Weight in gms	Tender Fruit Colour	Mature Fruit Colour
SC-6-0	15.6	17.6	4.2	9.4	Green	Light Green
SC-13	13	15.1	2	4.2	Green	Light Green
SCT-2 As	35.2	38	10	16	Deep Green	Deep Green
SCT-6	12.8	17.2	3	9.4	White Green	White Green
SCT-6-2	15.6	19	4.4	10.2	White Green	White Green
SCT-15-1	19.6	21	6.2	9.2	Green	Light Green
SCT-27-1	16.4	19	5	8.2	Light Green	White Green
SCT-27-4	17.2	21.6	5	10	Green	Light Green
SCT-34-4	21.6	27.2	6.2	11.6	Light Green	White Green
SCT-34-6	21.8	25.6	5.8	10.6	White Green	White Green
SCT-34-6-1	15.8	19.2	3.4	6.4	Green	White Green
SCT-34-6-2	19	19.8	5.6	6	White Green	Green
SC-130	14.4	17.2	4	7.8	Deep Green	Green